THERMAL EXPANSION 6

A Continuation Order Plan is available for this series. A continuation order will bring delivery of each new volume immediately upon publication. Volumes are billed only upon actual shipment. For further information please contact the publisher.

THERMAL EXPANSION 6

Edited by

Ian D. Peggs

Atomic Energy of Canada, Ltd.
Pinawa, Manitoba, Canada

PLENUM PRESS · NEW YORK AND LONDON

Library of Congress Cataloging in Publication Data

International Symposium on Thermal Expansion, 6th, Hecla Island, Man., 1977.
  Thermal expansion 6.

  Includes bibliographical references and index.
  1. Expansion (Heat) — Congresses. I. Peggs, Ian D. II. Title.
QC281.5.E9I57 1977   536′.41   78-11252
ISBN-13: 978-1-4615-9088-0   e-ISBN-13: 978-1-4615-9086-6
DOI: 10.1007/978-1-4615-9086-6

Proceedings of the Sixth International Symposium on Thermal Expansion
held on Hecla Island, Manitoba, Canada, August 29—31, 1977

Plenum Press, New York is a division of Plenum Publishing Corporation
227 West 17th Street, New York, N.Y. 10011

PREFACE

This 6th International Symposium on Thermal Expansion, the
first outside the USA, was held on August 29-31, 1977 at the
Gull Harbour Resort on Hecla Island, Manitoba, Canada. Symposium
Chairman was Ian D. Peggs, Atomic Energy of Canada Limited, and our
continuing sponsor was CINDAS/Purdue University.

We made considerable efforts to broaden the base this year to
include more users of expansion data but with little success. We
were successful, however, in establishing a session on liquids, an
area which is receiving more attention as a logical extension to
the high-speed thermophysical property measurements on materials at
temperatures close to their melting points.

The Symposium had good international representation but the
overall attendance was, disappointingly, relatively low. Nevertheless, this enhanced the informal atmosphere throughout the meeting
with a resultant frank exchange of information and ideas which all
attendees appreciated.

A totally new item this year was the presentation of a bursary
to assist an outstanding research student to attend the Symposium.
We were delighted to welcome Mr. Benedick Fraass from the University of Illinois to the Symposium, and he responded by making an
informal presentation on the topic of his research. We hope this
feature will continue.

Previous Symposia in the series were:

DATE	LOCATION	CHAIRMEN	SPONSOR
September 18-20 1968	Gaithersburg, Maryland	R.K. Kirby	Natl. Bureau of Standards
		P.S. Gaal	Westinghouse Astronuclear Lab.
June 10-12 1970	Santa Fe, New Mexico	R.O. Simmons	Materials Res. Lab. Univ. of Illinois
		D.C. Wallace	Sandia Laboratories

v

DATE	LOCATION	CHAIRMEN	SPONSOR
October 27-29 1971	Corning, New York	H. Hagy G.M. Graham	Corning Glass Works Univ. of Toronto
November 7-9 1973	Lake of the Ozarks, Missouri	R.E. Taylor G.L. Denman	Purdue University Air Force Materials Laboratory
June 4-6 1975	Storrs, Connecticut	T.G. Godfrey P.G. Klemens	Oak Ridge National Laboratory Univ. of Connecticut

The next Symposium will be held in 1979 in Chicago, chaired by
Dr. D. Larsen of IITRI.

I would like to express my thanks to Peter Gaal and the
members of the Governing Board for their support and encouragement
during the two years prior to the Symposium - Paul Wagner, Keith
Kirby, Gordon Godfrey, Bill Plummer, G. Ruffino, Ralph Simmons,
Guy White, and Ray Taylor. I much appreciate the work done by the
Session Chairmen: Prof. R.O. Simmons (University of Illinois),
Prof. G. Ruffino (Leeds & Northrup, Italy), Prof. P. Desré (CNRS,
France), Mr. M. Hammond (Boeing Canada), Dr. R. Roberts (CSIRO,
Australia), Prof. K. Krishna Rao (Osmania University, India),
Dr. P. Wagner (Los Alamos), and Dr. E.G. Wolff (Aerospace Corpora-
tion).

At the present time, there is much discussion on the need for
an umbrella Thermophysics Congress to coordinate the activities of
the several thermophysical-property-oriented organizations. My
experiences over the past two years indicate that such a group is
badly needed and I, for one, will give it full support.

I.D.P.

CONTENTS

APPENDIXES

REACTOR CORE MATERIALS FOR ADVANCED

HIGH-TEMPERATURE REACTOR SYSTEMS

H. Nickel

Kernforschungsanlage Jülich GmbH

Institut für Reaktorwerkstoffe, FRG

ABSTRACT

The advantage of the gas cooled high-temperature reactor (HTR) lies in its development potential for both the direct cycle gas turbine application and the generation of nuclear process heat at core gas outlet temperatures between 850 and 1000°C maximum. The change over from the steam cycle HTR, having a maximum core outlet temperature of 750°C, to the direct cycle or the nuclear process heat HTRs, having these higher gas temperatures, presents a considerable challenge to the material technology involved.
In this paper, problems arising in these new reactor applications and the present state of the art concerning the core-specific components are described. This group includes coated fuel particles, fuel elements for the different HTR-types (spherical elements and block-type elements), and the reflector.

INTRODUCTION

A study of world energy supplies reveals that during the past decade they have been characterized by a general abundance and a radical change in structure. In the period 1960-1970 the demand for solid fossil fuel increased by 24 %, while a rise of 100 % was registered for mineral oil and 117 % for natural gas. At this rate the mineral oil resources known today are likely to be used up by the beginning of the 1990s. The provision of further quantities of oil will be possible only by increased exploration, an improved petroleum displacement ratio or recovery of oil from oil sand and bituminous shale.[1]
The coal deposits in the Federal Republic of Germany and some other countries in Western Europe appear more or less adequate. It

1

is, however, not realistic - in absence of any important mineral oil or natural gas resources in the FRG - to cover the energy demand by increased conventional use of coal. The reasons are that (a) the cost of the coal from the Western European mines is high; (b) the consumption of coal in industrial countries at the present time depends on its use for the production of steam and electricity, for house heating etc.; and (c) the use of coal results in an increasing need for environmental protection as a result of higher ecological exposure (e.g. CO_2).

Increased use of coal will only be possible by building special plants, in which the coal is converted into gaseous and liquid energy carriers. These processes require a relatively high specific energy input. For the production of synthetic natural gas, for instance, approximately 45 % of the coal charge is required for the plant's power supply. The use of high temperature reactors for process heat applications provides a very promising alternative. The provision of low-cost nuclear process heat enables the application of reforming processes to fossil fuels.

HIGH-TEMPERATURE REACTORS (HTRs) FOR ADVANCED APPLICATIONS

The development of tne high-temperature reactor started in the mid 1950s. Since this time the technology of the system has proceeded steadily, being centred around the development of fission-product-retaining, ceramic fuel particles which are embedded in graphitic materials, the exploitation of an all-graphite core construction and the use of helium gas as coolant. Table 1 shows the existing variants of the HTR design.

Operation of the Dragon Reactor Experiment in the United Kingdom and of the test reactors - the Peach Bottom Reactor in the USA and the AVR in the FRG - has resulted in the construction of two prototype plants: the Fort St. Vrain with 330 MWe in the USA and the THTR (Thorium High Temperature Reactor) with about 300 MWe in the FRG. The design for a Prototype Reactor Nuclear Process Heat "PNP" in the FRG and different designs of HTR power-stations are being dicussed.

The HTR has already reached the stage of commercial exploitation in the field of electricity generation through the use of the steam cycle, which typically uses primary coolant helium at approximately $750^{\circ}C$. However, such temperatures in the primary coolant are by no means a limit in terms of the HTR core. One now sees the emergence of what one may term "the advanced high temperature reactors systems" where the outlet gas temperature will be raised and the heat from the coolant gas used in advanced applications such as HHT (High Temperature Reacotr with Helium Turbine) and PNP (Prototype Nuclear Process Heat Reactor).

Examples for application of nuclear process heat in endothermic reactions are:
-     gasification of coal either by hydrogenation or by using steam

- oil hydrocracking
- steam reforming
- chemical heat pipe
- water decomposition.

All these reactions require reactor coolant outlet temperatures between 850 and 1000°C, which appear feasible by further development.[2]

In Germany the nuclear process heat will be used for processing the fossil fuels. The incorporation of an HTR into the coal gasification process will be facilitated by the development of hydrogasification and steam gasification methods. Two non-nuclear test facilities on a semi-technical scale with a through-put of 100 and 200 kg coal/h,respectively,have become operational and will allow investigation of these processes.

First application of nuclear process heat from high-temperature reactors will be for the conversion of coal. Usage of nuclear heat

Table 1:   Existing HTR design variations

Experimental Reactors			
	Peach Bottom	Dragon	AVR
Operational	1966–1974	1968–1975	1968–
Thermal/electric power (MW)	115/40	20/ –	46/15
Fuel element type	prismatic	prismatic	spherical
He-inlet/outlet temperature (°C)	377/750	350/750	270/950
Mean He-pressure (bar)	25	20	10
Prototype Reactors			
	Fort St. Vrain	PNP	THTR
Operational	1976 (reduced power)	1982 (begin constr.)	1980 (target)
Thermal/electric power (MW)	843/330	750/–	750/300
Fuel element type	block	spherical	spherical
He-inlet/outlet temperature (°C)	405/784	300/950	270/750
Mean He-pressure (bar)	45	39	39
Commercial Power Reactors			
	HTR-K-1000		HTR-1160
Operational	earliest 1983		planning suspended
Thermal/electric power (MW)	2600/1000		3000/1160
Fuel element type	spherical		block
He-inlet/outlet temperature (°C)	275/750		320/730
Mean He-pressure (bar)	~ 50		50

will be for the production of process gas, chiefly in methane (synthetic natural gas - SNG), hydrogen and CO- and H_2-bearing synthesis gases as well as for nuclear energy transport systems (chemical heat pipe).

Great significance is to be attached to the production of synthetic gas ($CO+H_2$) in the steam reforming system for different sorts of light hydrocarbons - such as methane, natural gas, gasoline and refinery gases - with steam to hydrogen or a mixture of hydrogen and carbon monoxide. These synthetic gases are then used for the different processes such as hydrogasification, direct reduction of iron ore, ammonia synthesis, methanol synthesis, Fischer Tropsch synthesis, hydrocracking, chemical pipe, etc.

Since 1974 a primary coolant temperature of $950^{\circ}C$ is already being achieved in the experimental reactor AVR.[3] This is substantiated by the results so far obtained in developing a wide range of materials for the specific components of the core of high temperature reactors. Whilst the fuel elements and their components, particularly the coated fuel particles and the graphite as a construction and reflector material, are key factors in this respect, the use of nuclear process heat from HTRs also demands suitable materials for the "out of core" components, such as hot ducts, intermediate heat exchangers (IHX) for the process heat plants, methane reformer tubes, and coal gasification plants. For these components Fe-Ni-Cr and Ni-Cr alloys are envisaged.[4] In this paper the specific core components only will be discussed.

HTR FUEL ELEMENTS

Two different approaches have so far been made with regard to the core structure and arrangement of fuel elements in HTRs. While rod-type fuel elements were used in the Dragon and the Peach Bottom reactors, hexagonal block-type fuel elements are being used in the reactor design developed by the General Atomic Company. In their Fort St. Vrain prototype reactor such block elements are operating and are also considered for the advanced reactor systems.[5]

The reactor concept developed in the FRG uses spherical fuel elements (6 cm in diameter) in pebble beds to form the core.[6]

SPHERICAL FUEL ELEMENTS

The AVR test reactor is being operated with about 100,000 of these spherical fuel elements; the thorium high temperature prototype reactor (THTR), however, will have approximately 700,000 of these elements (Fig.1). The THTR specification values are: fuel element surface temperature $1000^{\circ}C$, coated particle maximum temperature $1150^{\circ}C$.

The quasi-isostatically pressed element developed by NUKEM/HOBEG Company, consists of a homogeneous composite system of a particle-matrix core and a fuel-free zone. This element exhibits an excellent

irradiation behaviour and a mechanical strength sufficiently high
for the use in a pebble bed reactor. One advantage of this type of
pressed fuel element is the relatively small particle volume loading,
12 to 16 %, ensuring a completely homogeneous envelopment of the
particles by matrix material. This results in a favourable, almost
spherically symmetrical, heat removal for each individual particle.
There is no gap between particle-containing zone and the fuel-free
zone of the spheres to impede further heat transfer directly towards
the cooling gas.

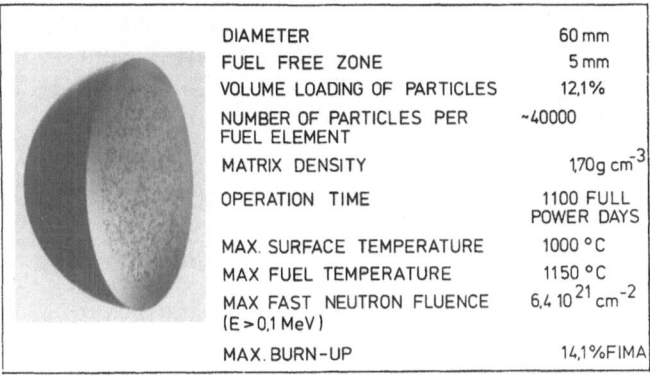

DIAMETER	60 mm
FUEL FREE ZONE	5 mm
VOLUME LOADING OF PARTICLES	12,1%
NUMBER OF PARTICLES PER FUEL ELEMENT	~40000
MATRIX DENSITY	$1,70 \text{g cm}^{-3}$
OPERATION TIME	1100 FULL POWER DAYS
MAX. SURFACE TEMPERATURE	1000 °C
MAX FUEL TEMPERATURE	1150 °C
MAX FAST NEUTRON FLUENCE (E > 0,1 MeV)	$6,4 \cdot 10^{21} \text{cm}^{-2}$
MAX. BURN-UP	14,1% FIMA

Figure 1:

Cross section and data
of the THTR-300 fuel
element

GAC BLOCK-TYPE FUEL ELEMENT

The GAC element consists of a hexagonal graphite block with
normally one coolant channel being hexagonally surrounded by 6 fuel
holes (Fig.2).

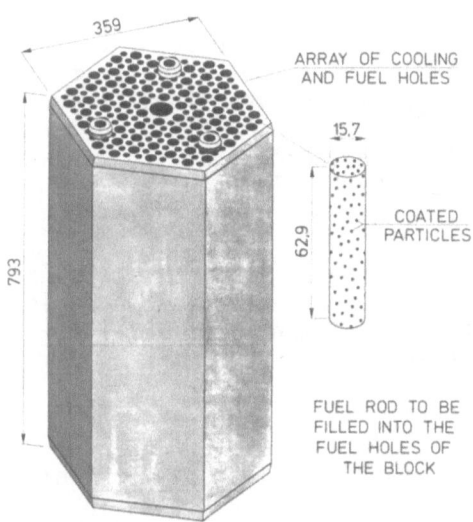

359

ARRAY OF COOLING
AND FUEL HOLES

15.7

COATED
PARTICLES

62.9

793

FUEL ROD TO BE
FILLED INTO THE
FUEL HOLES OF
THE BLOCK

Figure 2:

HTGR block fuel element

The fuel rods consist of a particle-matrix mixture with a particle volume lying between 20 and 55 %, graduated for the different core zones. Due to the relatively low thermal conductivity of the rod material and due to the gap between fuel rod and surrounding graphite, relatively high temperatures and temperature gradients may arise inside the particles. Possible damage to particles due to fission gas pressure, fast dose and chemical interaction between fuel kernel and PyC coating (amoeba effect) can be countered by an appropriate design of particle and fuel element, however.

During extensive irradiation experiments following the GAC design parameters, the fuel components - coated particles, fuel rods and, in particular, graphite - have been tested and good results were obtained. However, due to the size of the block-type elements, irradiation of fuel size blocks will only be possible in the Fort St. Vrain reactor.[5]

HTR FUEL ELEMENTS FOR ADVANCED REACTOR CONCEPTS

From the results mentioned about the present state of fuel element development in the FRG and the development potential of reactor systems with spherical or block-type fuel elements for achieving high gas temperatures one can summarize it as follows:

1) For commercial steam cycle HTR, the pressed spherical fuel element and the graphite block element following the GAC design have been successfully developed.

2) For advanced reactor systems HHT and NHP, comparing the potential of the two fuel element systems with regard to high helium temperatures, it can be stated for the spherical fuel element that performance of the present design as tested in irradiation experiments may incorporate 1250°C as the nominal fuel temperature and 1350°C as the peak random hot spot temperature. In a pebble bed reactor with OTTO fuelling (this means Once Through Then Out) and with a power density of 9 Mw/m^3,an increase of the gas temperature to 1190°C would be theoretically feasible (first of all without regard to the fission product release). This would entail a maximum nominal fuel temperature of 1240°C.

In the case of the GAC block reactor, an increase of the gas and fuel temperature should be also possible but would require a number of technical modifications compared to the fuel element for the steam cycle system.

Because of the excellent irradiation behaviour and good physical and mechanical properties of the spherical fuel elements the decision was taken in the FRG to use this spherical element type for the Prototype Reactor for Nuclear Process Heat.

COATED PARTICLES

Irrespective of the design, a feature common to all types of fuel elements is the fact that they contain the fissile and fertile materials in the form of coated particles (Fig.3). These spherical particles consist of a kernel of carbides or oxides of a uranium-thorium mixture or uranium and thorium respectively (about 200 to 600 μm diameter) and an effective fission product barrier, comprising several different pyrolytic carbon (PyC) layers (BISO) and, where necessary, an additional SiC-Layer (TRISO).

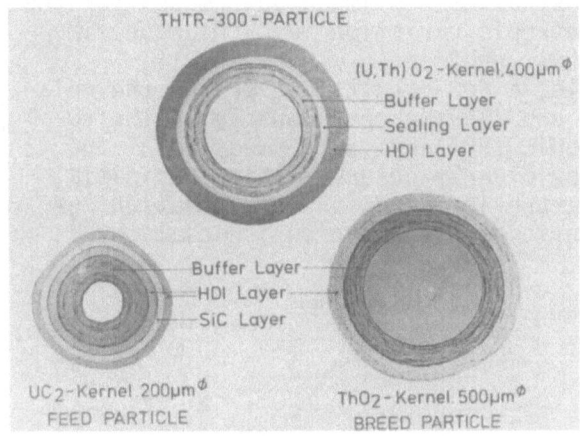

Figure 3:

Coated particles for U/Th-fuel cycle

For reasons of improved heat transfer, mechanical integrity and safety under reactor operating conditions, the coated particles are homogeneously dispersed in a graphite matrix both in the spherical and the block-type fuel elements and their irradiation performance has been shown to be satisfactory.[5-7]

Naturally the main objective of the coated particles is to produce heat in the uranium oxide or carbide fuel kernel. A further improtant requirement to be met is an optimum retention of fission products in the coated particles. Initial irradiation experiments resulted in the general observation that only those coated particles having isotropic layers (BAF values between 1.0 and 1.05) and PyC-densities between 1.75 and 1.9 g.cm^{-3} will have the necessary neutron irradiation resistance and, consequently,cope with the required service life in a power reactor. However, additional extensive irradiation experiments showed that these initial empirical data were not sufficient for the understanding of the influence of both burn-up and fast neutron fluence on the irradiation damage of the coating materials. At present, it is known that the life expectancy of the particles is essentially limited by two factors:
1) failure of the layers due to mechanical stresses and
2) chemical interaction of the oxide or carbide fuel kernel with the PyC coating (amoeba effect and/or the SiC corrosion).

Mechanical stresses produced during irradiation are caused by:
- internal gas pressure of fission gases and CO/CO_2,
- swelling of the fuel kernel and
- neutron-induced dimensional changes in the coating materials, for example,shrinkage and creep in the layers and interaction between fuel kernel and coating and between the layers themselves. Damage

that may arise from the internal gas pressure in the particle and
from the swelling of the fuel kernel may be largely compensated for
by application of a porous PyC-buffer layer.

On the other hand, it is substantially more difficult to eliminate
the neutron-induced dimensional changes which may lead to the failure
of layers as a result of high stresses. An improtant reason for the
formation of stresses is anisotropic variation in pyrocarbon dimensions under the influence of fast neutrons.

Fig.4a,b show, as an example, a few results concerning the relative change in the dimensions and the apparent density of the so-
called high temperature isotropic (HDI)-PyC as a function of the
fast neutron fluence. It can be clearly observed that the initial
density of pyrocarbon has a marked influence on the variations of
both linear dimensions and density due to neutron irradiation.[8]

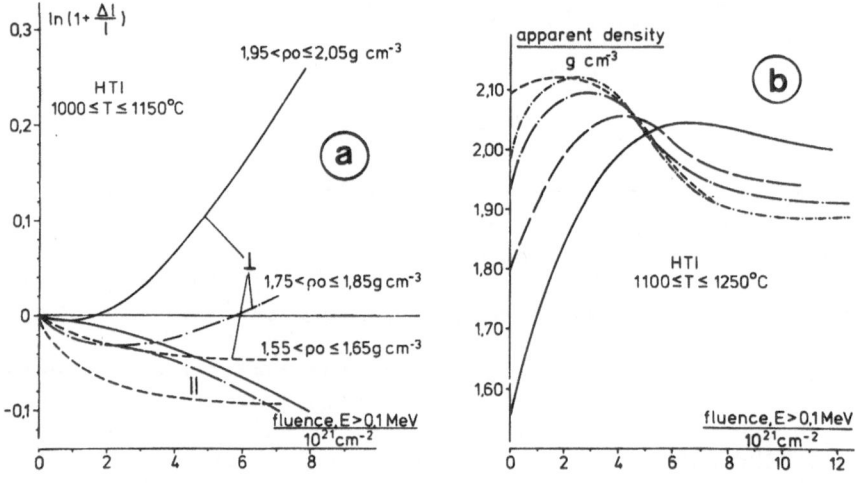

Fig. 4a, b: The effect on the fluence of fast neutrons on the
             properties of HTI-PyC:
             a) relative dimensional changes of PyC of different
                 densities irradiated at temperatures of
                 1000 to 1150°C
                 (perpendicular and parallel to the deposition plane)
             b) changes of the apparent densities of specimens[8]
                 irradiated at temperatures of 1100 to 1250°C

In order to determine the mechanisms of these damage processes
with a view to finding the most suitable of materials, some new
physical and chemical measurement procedures have been developed.
Only with the aid of these methods was it possible to make the
measurements,on the very small particles, needed to describe the
properties of the pyrocarbon and SiC-coating material (e.g. optical
anisotropy, crystallite size, microporosity, oxidation behaviour,
rupture strength, Young's modulus and creep). Understanding of the
mechanism of radiation damage will lead to the production of improved

materials. Certain requirements must be met if the further development of the steam cycle HTR to the advanced reactor system can be
achieved.

Microstructural examinations have shown that the pyrocarbon
deposited on the fuel kernels in fluidized beds is not homogeneous,
but is of composite structure.

This microstructure of the PyC can be determined by means of
transmission electron microscopy (TEM). TEM photomicrographs on
PyC layers of coated particles show that the microstructure of the
pyrocarbon consists of three components within a growth feature and
varies with temperature and concentration of the hydrocarbons used.
These three components within a growth feature are called "moasic"-,
"tangled"- and "layered" components. Recent results show irradiation
performance and fission product retention of pyrocarbon to depend
strongly on the type, volume concentration and on the distribution
of the microstructural components in the coating. For instance an
insufficient content of the tangled fiber component, which serves
as a buffer for the highly anisotropic dimensional changes of the
mosaic and the layered component, will lead to fracture of the pyrocarbon coating at even low neutron fluences. On the other hand, a
very high concentration of the tangled fiber component may result
in increased fission gas permeablility and a low preirradiation
fracture strength; the latter may cause coating failures during fuel
element production.

Current investigations are directed towards ensuring reproducible
production of optimised coatings in regard to irradiation stability
and fission product retention.

COATED PARTICLES FOR ADVANCED REACTOR CONCEPTS

Improved Fission Product Retention

The development of high-temperature reactors for use as direct-
cycle plants or for process heat production aims not only to reach
higher coolant temperatures, but, at the same time, to minimise contamination of the primary circuit. Thus, apart from safety-engineering
aspects, it should be possible at any time to carry out inspection
and, if necessary, repair of the primary circuit components. It is
therefore a general requirement to produce coated particles whose
fission product release rate will be one or two orders of magnitude
lower than that of particles used in fuel elements for steam cycle
systems. This is particularly true for solid fission products such
as strontium, barium and caesium, for which a pyrocarbon coating of
the particles in its present form will not offer a sufficient diffusion barrier. As a result of extensive irradiation tests on the
existing particle types with the specified burn-up and fast neutron
fluence level, it has been found that the maximum allowable particle
temperature in HTR fuel elements is of the order of 1350°C. Essentially, the release rate for solid fission products from these

coated feed-breed particles can only be reduced by the following
two methods:
1) utilization of coating materials with an improved retention
capacity for solid fission products, e.g. TRISO layers with SiC,
and
2) increased capacity of fuel kernel retention by kernel additives,
e.g. "$Al_2O_3+SiO_2$" alumina plus silica, which chemically form compounds with solid fission products.

Generally, it can be stated that, compared to PyC layers, SiC
layers in the investigated temprature range up to 1350^oC give a
nearly complete retention of the specified fission products.

The possiblility of retaining solid fission products in the fuel
kernel by adding certain high melting point oxides has already been
demonstrated in numerous out-of-pile and irradiation experiments.
At irradiation temperatures of $1100-1250^oC$ the release of the
fission products strontium and barium from coated particles could
be reduced by two orders of magnitude by addition of Al_2O_3/SiO_2
to both UO_2 and $(U,Th)O_2$-kernels. Using internal Cs release data
obtained from irradiation experiments, the effective diffusion
coefficient of caesium in the kernel at temperatures below 1200^oC
was found to be reduced by about two orders of magnitude, if the
kernels contained alumina-silica additives in amounts of only 5 wt.%.
This phenomenon works extremely well at the required temperature
range for coated particles in a process heat reactor with spherical
fuel elements.[9]

Amoeba Effect

The amoeba effect results from an attack of the coating layers by
a chemical reaction between kernel and PyC, the kernel moving in
the direction of the hot side of the particle. This effect is
observed irrespective of the particle design, U-Th- or uranium low
enriched-cycles, carbide or oxide kernel, and structure of layer.
The amoeba effect is due to the asymmetric temperature distribution
in the fuel particle at high temperatures.

In the case of coated particles with a carbide kernel, carbon
transport is observed in the solid phase as a function of the
temperature gradient in the carbide phase.

With oxide particles, the mechanism has not yet been fully
evaluated. However, it is well known that the chemical potential
of the oxygen is of major importance for carbon transport. For
process heat reactors, the problem of avoiding the amoeba effect
will be of great significance, especially for HTRs with block-type
elements. A method of avoiding the amoeba effect exists by reducing
the oxygen potential using a slight addition of UC to the UO_2 kernel.
This notably reduces the rate of carbon transport, which greatly
influences the amoeba effect.[10]

FUEL ELEMENT MATRIX AND GRAPHITE

Graphite and the graphitic matrix form the structural materials of
the HTR fuel elements. Apart from their cross section and moderating
properties they are selected primarily because of their mechanical
properties under reactor conditions, their production technology
and availability and, to a lesser extent, because of their ability
to retain fission products by absorption. Extensive knowledge of the
behaviour of these materials under HTR conditions has been, and is
being, obtained by extensive irradiation testing.[5,11,12] These
tests also indicate the trend of further developments towards higher
coolant outlet temperatures. Here only some aspects of the thermal
expansion of graphitic materials are discussed.

THERMAL EXPANSION OF GRAPHITIC MATERIALS

Amongst the properties which make graphitic materials suitable for
the application in HTRs, thermal expansion plays an important role.
The Coefficient of Thermal Expansion, CTE, is low compared with
those of other materials, but there are significant differences
between various graphite types so that the values for the CTE can
vary by a factor of 4. Therefore the knowledge of the CTE with
respect to the particular graphite type is of fundamental interest.
Though under normal reactor conditions the thermal stresses are
reduced by irradiation induced creep, they are built up again with
reverse sign during reactor shut-down, because the crystallites
which had expanded during heating up contract during cooling down.
By this contraction, tensile stresses are generated which can not
be lowered by thermal creep, because this occurs only at temperatures
higher than 1800°C. For the reduction of the inner stresses
mentioned above a low CTE is advantageous. It should not be greater
than $6.10^{-6}/°C$ measured between 20 and 500°C. The value for the
CTE depends upon the kinds and amounts of the raw materials used
as well as upon the forming procedure and heat treatment.
     Figs.5 and 6 may give a certain information about the influence
of the production parameters on the value for the CTE.

An important production parameter of graphitic materials is their
binder content. Okada has shown that with increasing binder content
the CTE also increases up to a maximum in the range of about 20
weight-% parallel and about 30 weight-% perpendicular to grain[15]
orientation and then decreases with increasing binder content.
The influence of binder content on the CTE of graphite matrix which
is used for the production of spherical HTR fuel elements is shown
in Fig.7. The final heat treatment temperature for these materials
is 1,800°C because of the coated fuel particles embedded in the
matrix.

Under fast neutron irradiation, the CTEs of graphitic materials
change significantly. There is an interrelation between these
changes and the irradiation induced dimensional changes. Both depend

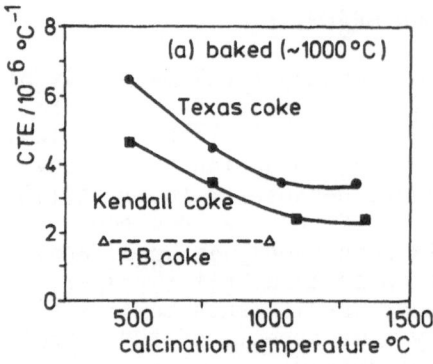 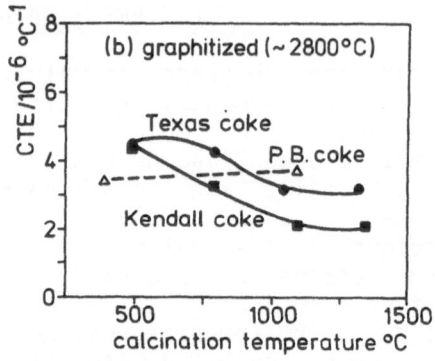

Fig. 5:   CTE perpendicular to grain orientation as a function
          of calcination temperature of the coke[13]
          a) after baking,   b) after graphitization

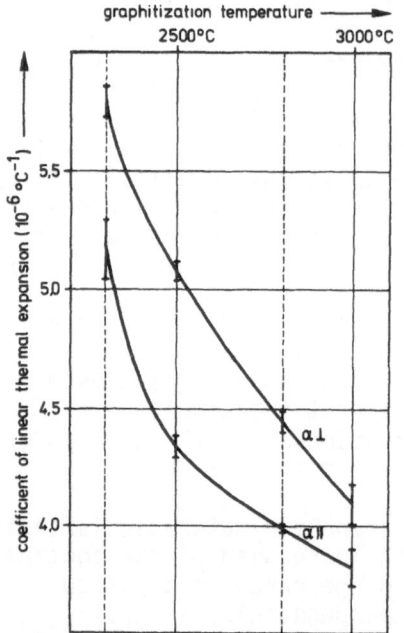

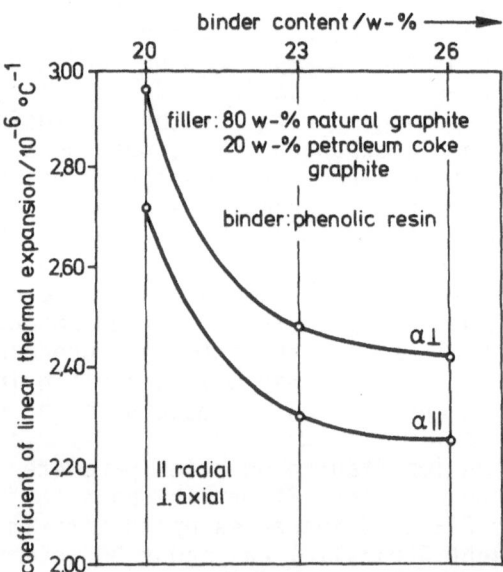

Fig. 6:
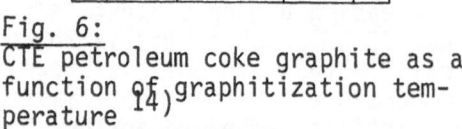
CTE petroleum coke graphite as a
function of graphitization tem-
perature [14]
⊥   CTE perpendicular to grain
    orientation
"   CTE parallel to grain orien-
    tation

Fig. 7:
Change of CTE with increasing
binder content for graphitic
matrix materials

upon the availability of micropores for buffering the thermal expansion of the graphite lattice.

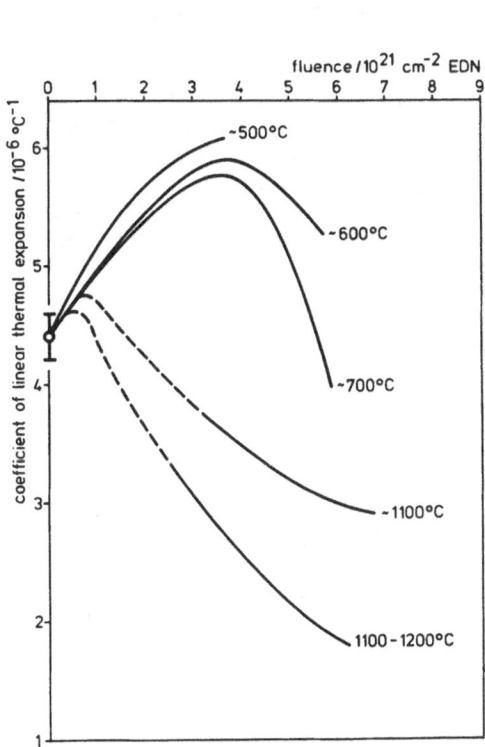

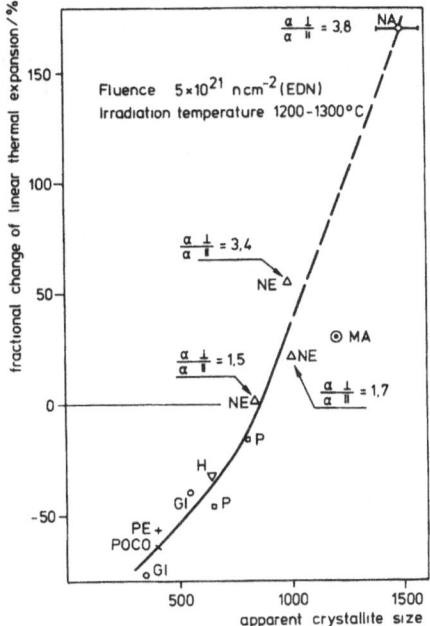

GI Gilsonite coke graphite
MA Graphite matrix 80 w-% natural graphite
                    20 w-% petroleum coke graphite
NE Needle coke graphite
P   Petroleum coke graphite
PE Pitch coke graphite
NA Natural graphite

Fig. 8:
Change of CTE as a function of fast neutron fluence for irradiation temperatures between 500 and 1200°C.[14]

Fig. 9:
Fractional change of CTE as a function of crystallite size.[16]

The CTEs of most materials increase under neutron irradiation because of the lattice expansion in the c-direction, which is a consequence of the formation of interstitial atoms and clusters.

The crystallites expand into the micropores, which are therefore more and more closed. When the temperature increases, the thermally induced lattice exapnsion can no longer be buffered by the micropores. Therefore an expansion of the grains - and to a smaller extent - of the graphite block occurs. When at higher fluences a great number of vacancies and vacancy clusters have been formed, a contraction of the lattice in the a-direction occurs which may lead to a decrease of the CTE of the graphite depending upon crystallite size and irradiation temperature as shown in figs.8 and 9.

The material which is discussed in fig.8 is isostatically pressed
petroleum coke graphite. The CTEs are equal for all orientation
directions. The influence of the anisotropy ratio on the irradiation
induced change of CTE can be seen in fig.9; the higher the anisotropy
the greater the change of CTE.

Fig.10 demonstrates the differences in irradiation behaviour (CTE)
as a consequence of the use of various cokes.

The influence of the graphitization temperature on the change of
the CTE is demonstrated in fig.11. The four graphites investigated,
consisting of the same raw materials, were baked at the same temperature but graphitized at 2300, 2500, 2800 and 3000 °C. The increase
in CTE observed in all other cases does not occur in the material
which was not fully graphitized. It should be mentioned that there
are also significant differences in the irradiation induced dimensional changes of the four graphites.

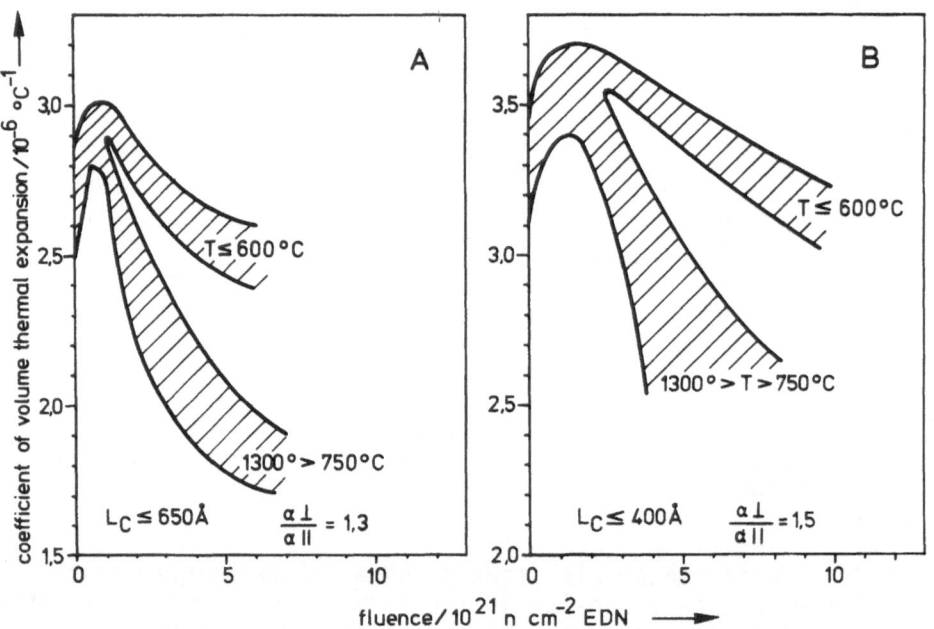

Fig. 10:
CTE of petroleum coke graphite (A) and pitch coke graphite (B) as
a function of fast neutron fluence for different irradiation
temperatures.[16]
L_C: Crystallite size in C-direction

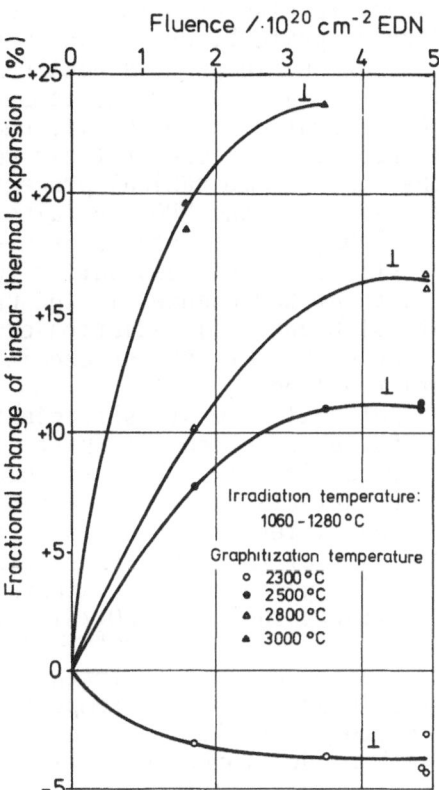

Fig. 11: Fractional change of CTE as a function of fast neutron fluence for petroleum coke graphites graphitized at different temperatures.[14]

SUMMARY

In the advanced high temperature reactor systems with direct cycle or for nuclear process heat application the core gas outlet temperatures will range from 850°C to 1000°C maximum. For the steam cycle HTR confidence in the materials has been gained by previous developments and excellent sample behaviour in testing, whereas for the advanced HTRs a considerable development effort is still required

'•ith a substantial material program on the way.

For commercial steam cycle HTRs both the pressed spherical fuel element and the machined graphite block element were brought to the point of technical application. In process heat reactors the gas outlet temperature rises from about 750°C to temperatures between 950° and 1000°C, and these systems also require reduced release rates for the fission products. Because of their good irradiation behaviour and their physical and mechanical properties, the spherical fuel elements are favored in the FRG, and also for Nuclear Process Heat. The comprehensive test results from steam cycle irradiations allow the statement that gas outlet tempratures up to 970°C may be attained without any changes in the present concept of spherical fuel elements although no irradiation experiments were mounted simulating the process heat reactor conditions with OTTO (Once-Through-Then-Out) fuelling.

For the present type of fuel elements sufficiently low fission product release rates can be expected with particles of mixed oxide kernels and a BISO coating. Improvements regarding fission product retention in the particles will be achieved by introducing TRISO-coated particles (particles with a SiC sandwich layer between the pyrocarbon). Another way of reducing the release rates lies in the bonding of solid fission products as for instance caesium and strontium to kernel admixtures such as Al_2O_3/SiO_2 in the fuel kernel.

A substantial investigation program is intended to answer open questions concerning the irradiation- and temperature-dependent damage mechanisms for coated particles. For this purpose, characterization methods have been developed that allow one to determine the various properties of kernel and layer materials prior to and after irradiation and, consequently, to characterize and to specify the materials. In this manner, it will be possible to determine the material changes under reactor conditions, thus showing methods for improvements in materials development.

Graphite and a graphitic matrix are used as structural materials for the HTR fuel elements and for the reflector. Knowledge about the behaviour of these materials under advanced HTR conditions have been, and is being, obtained by extensive irradiation testing.

For structural graphite the present situation is characterized by the effort in developing and testing of graphite for the highly stressed reflector of the pebble bed reactor. Test results obtained so far confirm that a suitable reflector material on a pitch coke base will be available in time and will repalce the earlier, fully tested gilsonite candidate materials no longer available on the market.

REFERENCES

1. G. Dietrich, H.G. Eickhoff, F. Niehaus, H.F. Niessen: Nucl. Eng. Des. 34, 1975, 3-13

2. V. Maly, R. Schulten, E. Teuchert: Atomwirtschaft 19, 1974, 601-603

3. W. Cautius, J. Engelhard, G. Ivens: Atomwirtschaft 19, 1974, 444-450

4. H. Nickel: High Temp. - High Press. 8 (1976) 123-141, and Proc. ANS-Conference on World Nuclear Power, Washington D.C. 1976, Transactions Vol. 24 (1976) 160

5. T.G. Gulden, J.L. Scott, C. Moreau: Proc. ANS Conference on Gas-Cooled Reactors, Gatlinburg, Tenn., 1974, Conf. 740501, 176-182

6. H. Nickel, E. Balthesen: Kerntechnik 17 (1975) 205-214

7. D.P. Harmon, C.B. Scott: J. Nucl. Techn. 1977 (in press)

8. J.C. Bokros, K. Koyama: J. App. Phys. 41 (1970) 2146-2155

9. R. Förthmann, E. Groos, H. Grübmeier: KFA-Report Jül-1226 (1975) 1-29

10. A. Naoumidis: J. Nucl. Mat. 48 (1973) 118-124

11. G.B. Engle, M.R. Everett, W.P. Eatherly: Proc. ANS-Conference on Gas-Cooled Reactors, Gatlinburg (Tenn.), 1974), Conf. 740501, 288-305

12. D.L. Leushacke, H. Cords, J. Semmler, E. Spann: Proc. IAEA Symp. on Gas-Cooled Reactors with Emphasis on Advanced Systems, Jülich/FRG, 1976, paper SM-200/33, Vol. I, 291-307

13. F.M. Collins: Proc. 1st Carbon Conf., pp. 177-187 (1956) Pergamon Press

14. J. Fourré, A. Pernollet, W. Delle, G. Haag: Preprints Int. Carbon Conf. CARBON '76, Baden-Baden, pp. 353-356 (1976)

15. J. Okada: Proc. 4th Carbon Conf., pp. 547-552 (1959) Pergamon Press

16. W. Delle: Preprints Int. Carbon Conf. CARBON '72, Baden-Baden, pp. 176-178 (1972)

THERMOPHYSICAL PROPERTIES: SOME EXPERIENCES IN RESEARCH, DEVELOP-

MENT AND APPLICATIONS

Paul Wagner, Los Alamos Scientific Laboratory
University of California

Russell U. Acton
Sandia Laboratories

INTRODUCTION

Over the years the Conferences on thermal expansion and thermal conductivity have been properly preoccupied with discussions of the theory and measurment of thermophysical properties. Actually, some of the uses to which this information has been put are of considerable interest in themselves, and some of the seemingly routine requests for thermophysical property information by workers in the applications area have led to complex, and sometimes basic research projects. Just for example, during the years 1955 to 1973, the Los Alamos Scientific Laboratory designed, built and tested thirteen graphite-core, gas-cooled nuclear reactors with the stated purpose of developing a nuclear rocket. The principle of operation was simple, the graphite core was heated and the energy transferred to the flowing gas propellant. Since thermal gradients were generated in the graphite core, stresses were generated. In general, stress generated by thermal gradients = $f(\alpha,E,Q,\lambda,X_i,\nu)$, where α is the thermal expansion coefficient, λ the thermal conductivity, and E, Q, X_i, and ν are Young's modulus, heat flux, dimensions, and Poisson's rate respectively. Since energy generation rates were so enormous[1] (a reactor core about the size of a 55 gallon oil drum could generate energy more rapidly than Hoover Dam), thermal gradients were very high and resultant stresses were very high. For proper core design, heat-up, control and test operation it was essential that λ and α be known as functions of temperature (to 2500°C) as accurately as possible. In addition to using this information as an analytical tool, maintenance of controlled values of λ and α in fuel rods dictated that we use these thermophysical properties for production control criteria. The anisotropic nature of graphites used added an additional complication to the measurements and techniques had to be devised [2-4] to overcome the influence of these directional characteristics.

This paper describes just a few of the paths that were followed as a result of involvement in some of the programs at the Los Alamos Scientific Laboratory and at Sandia Laboratories.

THE NUCLEAR ROCKET PROGRAM

In addition to the material discussed in the Introduction, there was an intimate coupling of the thermal expansion characteristics of the graphite core and the reactor control in the nuclear rocket engine. As the temperature was increased the graphite and uranium expanded, this decreased the overall core density, and therefore the ability of the core to moderate neutrons (and maintain the required excess reactivity). This, in turn, meant that as the reactor operating temperature increased the control drums in the reactor had to allow more neutrons to reflect back into the core. The rule of thumb used was that the decrease in the density of the graphite caused by heating to 2500°C decreased the excess reactivity that was used for control by 15-20%. Since the control margin wasn't all that large to start with, a very rigorous testing and production control schedule designed to monitor the CTE was instituted at LASL and at contract fuel rod fabrication centers. The most elaborate of these was built at the Westinghouse Astro Nuclear Laboratory.[5] This facility could make research grade measurements simultaneously on ten fuel samples from room temperature to 2500°C. Both graphite and uranium-containing graphite fuel rods were tested in this CTE apparatus as a routine portion of the fabrication program.

At LASL, we actually monitored the overall expansion of full length (137 cm) fuel rods to ensure that the sampling procedure for CTE samples was indeed representative of the entire rod. This non-destructive test was done by measuring the change in length produced when a quartz-encased fuel rod was immersed in an isothermal oil bath at 150°C. The expanding rod moved a push rod which contacted a dial gauge. These measurements were very sensitive and proved to be very accurate when performed carefully. For our purposes, this was the ultimate demonstration that measurements made on small, appropriately sampled CTE specimens yielded results with the liquid accuracy.

GRAPHITE RESEARCH

As an adjunct to the nuclear rocket program, LASL funded a research and development effort whose goal was to study the structure of graphite and to understand and fabricate predictable and reproducible graphites. One of the most important data inputs in the evaluation of experimental materials proved to be results obtained from

measurements of linear coefficient of thermal expansion (CTE). The
reasons for this are three-fold. First, the measurement of CTE is
by its nature an averaging type measurement which is independent of
specimen shapes and is not affected by fabrication or machining flaws.
Second, by virtue of its structure, graphite is highly anisotropic
and a rapid and convenient measure of artifact anisotropy may be obtained from CTE measurements. Third, the absolute value of the CTE
must be considered for practical applications and uses of the graphites (see the previous section). Crystalline graphite has a layered
structure with a C-C distance of 2.46 A and an interlayer spacing of
3.35 A. This large interlayer spacing is manifested in a much larger
CTE than the in-plane ("a") value: for example at about 800°C the
"c" direction CTE is about $8 \times 10^{-6}/$°C while the "a" direction CTE is
approximately zero. Furthermore, between 25°C and about 400°C the
in-plane expansion is actually negative. In a polycrystalline graphite the thermal expansion properties will depend on the spatial
arrangement of the crystallites which make up the material or, turn-
ing this around, measurements of CTE yield important information on
the distribution of crystallite "c" axes. Of utmost importance is
the actual value of the CTE in graphite since its low value contributes to the thermal stress resistance of this material. On the
other hand, a low value of the CTE in graphite is a disadvantage
when used with metals or with metal carbide systems since the latter
have relatively high CTE's and the mismatch may cause severe mechanical problems. The CTE of graphite can vary widely with raw materials
and manufacturing procedures and means of tailoring the CTE of the
final product improved rapidly as our understanding of the structure
of polycrystalline graphite increased.

Another spin-off from the graphite research program was the
research done on boron-doped, artificial, polycrystalline graphites,
[6,7]. The original intent was to tailor the electrical properties of
the graphite by adding small amounts of boron. Indeed our first
measurements showed an increase in the electrical conductivity of
about 35% upon addition of 0.3 percent B. X-ray diffraction studies
on these materials showed a marked decrease in the d_{002} (i.e. interlayer) spacing with increased B content. We concluded (as had others)
that at low B concentration studied, the boron atoms occupied sub-
stitutional sites in the lattice and as such would be expected to en-
hance the electrical conductivity in the in-plane direction.

We reasoned that the observed decrease in d_{002} was a reflection
of an increase in the interlayer ("c" direction) bonding forces and
should affect the thermal expansion characteristics. Since the out-
of-plane vibrational modes are more readily excited than the in-plane
modes, the thermal expansion in this, the "c" direction is much
greater than it is in the direction parallel to the layer planes. Because of this large anisotropy in the polycrystalline system, the
measurement of gross thermal expansion (especially at low temperatures)
in any particular direction will, to a large extent, be a measure of

the "c" direction component. From the d_{002} data, we anticipated a
decrease in the "c" direction component. We then chose samples of
identical anisotropy (so that the crystalline effects would not be
masked by anisotropic effects) and measured α from $23°$ to $645°C$. We
did indeed find a monotonic decrease in the average CTE of about 20%
with B content (up to 0.8% B). This was further confirmed by noting
an increase in Young's modulus with increased boron content up to the
solubility limit.

Since we noted an increase in the electrical conductivity (σ)
of the graphite with increased boron content and since there has been
a long-standing challenge to the experimenter to demonstrate unambig-
iously, the fractional value of lattice thermal conductivity component
in graphite,[8] we also made measurements (at $23°C$) of thermal conductivity. Since

$$\lambda = \lambda_{electronic} + \lambda_{lattice}$$

and by the Weidmann-Franz law

$$\lambda = L\sigma T + \lambda_{lattice}$$

we felt we could extrapolate σ back to zero and get the lattice component. Unfortunately, λ decreased with increased B content. Appar-
ently the B atoms act as phonon scatters and $\lambda_{lattice}$ decreases more
rapidly than $\lambda_{electronic}$ increases as the boron atom concentration
increases; however, the magnitude of the decrease in λ exceeds that
predicted by simple scattering theory[7,9] and a truly satisfactory
explanation of this phenomenon has never been published.

THE SNAP 19/PIONEER HEAT SOURCE INSULATORS

LASL was asked to look at the CTE and thermal cycling behaviour
of the three nested, pyrolytic graphite tubes which comprised the
SNAP 19/Pioneer heat source insulator. The source of concern was that
the insulators would break down in some manner upon thermal cycling
and fail the reliability tests. As it turned out,[10] the cylindrical
insulators distorted and also grew in length with repeated thermal
cycling between room temperature and $2500°C$. X-ray diffraction measurements made before and after heating showed an increase in the
crystallite height, a decrease in the interlayer spacing, and an increase in the orientation factor. All this indicated an overall increase in the crystallinity of the pyrolytic graphite upon heating.
It was found that high temperature annealing the pyrolytic graphite
in a constrained configuration and remachining produced a thermally
stable insulator.

SANDIA GRAPHITES

During the years 1969 to 1975, Sandia Laboratories designed, built and flew a number of high performance re-entry bodies using graphite nose tips, heat shields and heat shield insulators. The kinetic energy of a re-entering body is transformed into heat, with surface temperatures approaching the sublimation point. Thermal gradients are thus generated in the re-entry bodies which produced stresses, as was discussed earlier.

Thermal properties govern more than thermal stresses within the RVs. As the graphite heat shield heats up and expands, the payload slips forward, changing the center of gravity and consequently, the flight characteristics of the vehicle. Before impact, however, the heat shield begins to cool, putting the insulation in compression. The insulation strength must be able to support the shield ambient and heatup conditions, but be crushable during the shrinkage phase in order not to fracture the heat shield. The graphite materials development program associated with these re-entry bodies was so closely coupled to the thermophysics laboratory that often property data was awaited in order to select the next change in a process variable. This close coupling necessitated the development of a properties laboratory that could generate data on large numbers of samples to high temperatures in short times. The thermal diffusivity apparatus that was designed as a part of that program is believed to still be unique--it is capable of measuring the thermal diffusivity of 20 samples from room temperature to 2500 K in an eight-hour day. The apparatus features computer data acquisition and reduction.

Thermal expansion measurements on these graphites were made in dilatometers. We experimented with multi-sample dilatometers but settled on single dilatometers with multiple furnaces. Expansion measurements were made at 20C per minute. Comparison tests at 2 and 5°C per minute showed no significant differences. At the end of a test, the hot furnace was removed from the dilatometer. Without the furnace, the sample tube and pushrod very quickly cooled to room temperature. A second "cold" furnace was placed on the dilatometer. In this fashion, three samples could be run on each dilatometer in an eight-hour day.

In conjunction with LASL, Sandia has developed a graphite nuclear fuel for advanced pulsed reactors. The same types of problems occur as was discussed earlier.

The thermal properties of the SNAP 19/Pioneer heat source insulators were also investigated by Sandia. The interesting aspect of this work was in the accident analysis. The thermal expansion of the close fitting sleeves of pyrolytic graphite cause the heat transfer rate to be different depending on the direction of heat flow. The heat flow is outward under normal operating conditions, but could be inward under accident or planetary re-entry conditions. The insulator thus acts somewhat as a thermal diode.

ENERGY RELATED PROJECTS

As in the case with many other laboratories, much of our work in the last three or four years has been on energy related projects. At some point in any energy program, the production, movement or storage of heat must be considered, thus the need for thermophysical properties. Most of the properties data in these areas is generated for use in mathematical models which attempt to describe the entire system and/or process.

LMFBR

The liquid metal fast breeder reactor is a prime candidate to meet the world's future energy needs. Safety considerations are the major stumbling block of this energy program. We are currently engaged in studying the case wherein the pressure vessel is breached and molten sodium, stainless steel and fuel come in contact with the concrete of the containment vessel. A knowledge of the thermophysical properties of the various components is essential in modeling these reactions, penetration rates, etc.

WASTE ISOLATION

Isolation is the terminology now being used in conjunction with radioactive wastes. The word disposal implies no further accountability or control, whereas isolation means just the opposite. Should nuclear reactors become the energy source of the immediate future, then waste isolation becomes a major problem.

Salt beds are being given prime consideration as radioactive waste isolation sites for a number of reasons:

(1) Salt beds are tectonically stable,

(2) No ground water problems (salt is water soluble. These beds have been around for about two million years. If there was any water around, they wouldn't be),

(3) Salt transports about twice as much heat as would granite. Waste storage containers would thus be cooler,

(4) Salt is plastic. Cracks and fissures tend to heal. Items placed in salt become imbedded with time, and

(5)   The mining and maintenance of a waste isolation facility
would be considerly cheaper in salt than in granite.

Thermophysical properties are necessary in the analysis
of the transfer of heat from the waste to the salt surroundings.

INSITU COAL GASIFICATION

The burning of a coal seam, in place in the ground, to produce a fuel gas, offers large energy savings and would allow the
full utilization of high sulfur coals. The thermal modeling of
this process is under investigation. Several burns have been
successfully conducted at Hanna, Wyoming, by the Laramie Energy
Center with Sandia Laboratories' assistance in the areas of instrumentation and data acquisition and analysis. Down hole instrumentation was thermally coupled to the surrounding strata through the
use of matching grouts. The thermal properties of the coal seams,
over and under burden and grouts were necessary to select the proper
grout. The instrumentation was thermally coupled so that the temperature field and burn front would be distorted as little as possible.

SUMMARY

Accurate measurements of thermophysical properties, understanding exactly any use requirements, and understanding the behaviour of
the material are all potential ingredients for in-depth research and
development projects. The work reported here is but a brief description of some of the many LASL and Sandia projects and their
chain-branching manifestations. Unfortunately, there is a wealth of
information on projects that have been pursued in the past 25 years
of the general type described here which are described in detail in
internal memoranda or informal reports and almost certainly will not
ever be published in the open literature.

REFERENCES

[1]   Spence, R.W., Sci. Technol. No. 43 59 (1965).

[2]   Wagner, P., and Dauelsberg, L.B., Carbon 5 271 (1967).

[3]   Wagner, P., and Dauelsberg, L.B., Carbon 6 373 (1968).

[4]   Wagner, P., Brit. J. Appl. Phys., 2 293 (1969).

[5]   Gaol, P.S., Thermal Expansion of Solids Symposium, Sept. 18-20,
1968 (Gaithersburg, MD).

6   Wagner, P., and Dickinson, J.M., Carbon $\underline{8}$ 313 (1970).

7   Kelly, B.T., and Tobin, D., Whittaker, A., and Wagner, P.,
     Carbon $\underline{9}$ 447 (1971).

8   Klemens, P.G., 4th Conference on Thermal Conductivity,
     October 13-16, 1964 (San Francisco, CA).

9   Klemens, P.G., private communications.

10   Wagner, P. unpublished work.

MATERIALS FOR LARGE SPACE OPTICS

F. Ayer, E. G. Wolff and G. G. Comisar

C. S. Draper Laboratory & Aerospace Corporation

Cambridge, MA & El Segundo, CA

ABSTRACT

Materials are compared on the basis of their ability to minimize static, thermal and dynamic load induced deflections at the center of a large spherical mirror segment or panel. The analysis considers a 0.5m diameter mirror in two environments expected in space applications: (A) low incident radiation and low operating temperature (150°K) and (B) high incident radiation, moderate operating temperature (350°K). Although the analysis permits examination of various geometrical and loading parameters, the chief objective has been to compare twelve promising mirror construction materials, including composites, glasses, ceramics and alloys. Fused silica and carbon/graphite are prime candidates for type A and B environments, respectively. The ranking of the materials is sensitive to local variations in thermal expansion coefficients, possible actuator compensation for thickness changes and/or improvements in properties perpendicular to the mirror plane. Additional considerations in the choice of mirror materials are also discussed.

INTRODUCTION

The design and performance of large optics structures in space represents an area of increasing technical activity. Detection, analysis and reflection of earth, solar and space radiation are exemplified by the large space telescope project or solar energy transfer concepts. Low beam dispersion and

increased power assure expanded use of lasers in space. Typical
applications include Nd - YAG lasers for earth sciences (e.g.
LAGEOS Satellite) and CO_2 lasers for communications (e.g. the
NASA Tracking and Data Relay System planned for the 1980's).

The performance of a structure is usually estimated from
the equations expressing the distortions as a function of the
structural characteristics and the expected loadings. The
behaviour of a mirror is, for simplicity here, associated with
the central deflection of sag δ of a circular thin plate with an
initial curvature of radius R. It is assumed simply supported
along its periphery (Figure 1). The mirror sag δ is affected by

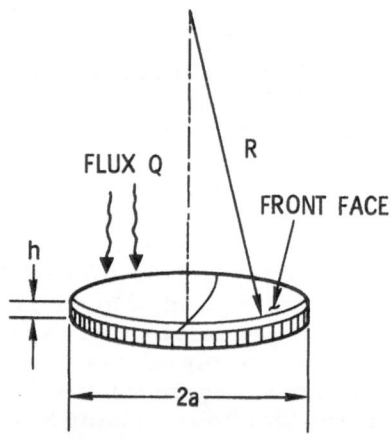

Figure 1. Mirror Description

the three following contributions, 1) geometry (F_G), 2) material properties (F_M), and 3) loading condition (F_L), in an expression of the form

$$\delta = F_G \cdot F_M \cdot F_L$$

The purpose of this discussion is to uncover those material properties (entering the equation through F_M) which affect most adversely a mirror exposed to the various loading conditions expected in space applications. In order to allow easier comparisons between materials it is desirable to keep the diameter (2a) and the curvature (1/R) of the mirror constant. In addition the equations are expressed in terms of constant mass per unit area (m) rather than constant mirror thickness (h) thus permitting a comparison which does not involve the launching cost factor. In the expression for δ, the term F_G resulting from geometrical data only is kept constant. Through F_L, the mirrors are subjected to various environments (thermal and dynamic) most likely to be encountered in space applications and some other loadings associated with gravity release and orbit injection. A survey of the basic equations covering these cases is presented.

The equations express deflections in terms of the structural rigidity of the plate but neglect, for simplicity, the effect of shear deformability. The shear component of the deformation is expected to be small for large span to thickness ratios of solid plates but will be significant in any sandwich type cross section.

MIRROR DEFLECTION EQUATIONS

Static Loading

The central deflection δ of a circular plate due to its own weight can be expressed in terms of its mass per unit area as

$$\delta \cong \frac{3}{4} a^4 \frac{(1-\nu^2)\rho^3}{E_\parallel} \left(\frac{I_s}{I}\right) \frac{g}{m^2} \qquad (1)$$

where

$E_\parallel$	=	Modulus of elasticity of plate material parallel to mirror plane
ρ	=	Density of plate material
ν	=	Poisson's ratio of plate material
g	=	Earth gravitational acceleration
I_s	=	Moment of inertia of solid section
I	=	Moment of inertia of built-up section of same area

Thus in order to optimize the deflection δ the material choice must minimize the parameter $(1-\nu^2)\rho^3 E^{-1}$. This case is primarily related to variations of gravitational fields (e.g. gravity release) and sometimes polishing loadings. We note that a small ratio I_s/I suggests that the use of a built-up or sandwich section requiring the same amount of material can improve δ substantially. It is understood that any sizable mirror will naturally require this kind of section.

Thermal Loading

The sudden application of a thermal absorbed flux Q to a structure causes transient effects followed, after some time, by a thermal equilibrium defined as steady state. The thermal state resulting from the latter is conveniently decomposed for analysis into a uniform temperature rise and a linear thermal gradient through the mirror thickness. The following equations describe these thermal conditions.

Uniform Rise Due to Incident Flux. Assume that the mirror is exposed to a thermal flux on its front face while its back face is interacting, through a grey body radiation coupling factor F_e, with a back-up structure at temperature T_o. Assume further that there is no re-radiation of the incident flux from the front face and that the back face temperature T_b is approximately equal to T_o. (If $T_b \sim 1.07\,T_o$, the error from the latter assumption is $\sim 10\%$). The central deflection resulting from the uniform temperature rise of $(1/2(T_f + T_b) - T_o)$ is

$$\delta = \frac{-a^2 \alpha_\parallel}{2R} \left[\frac{m}{2K_\perp \rho} + \frac{1}{4\sigma T_o^3 F_\epsilon} \right] Q \qquad (2)$$

where

$\alpha_\parallel$ = Coefficient of thermal expansion parallel to mirror plane

$K_\perp$ = Thermal conductivity of mirror material perpendicular to the mirror plane

σ = Stefan-Boltzmann constant

In this expression the material parameter α/K is likely to be predominant. It is of interest to note that a flat mirror will not deflect under a uniform rise since R is infinite, for this case.

Linear Gradient. Since the top and bottom fibers of the mirror cross-section experience different temperatures, unequal extensions cause the plate to bend. The central deflection expressed in terms of the absorbed flux Q is

$$\delta = - \frac{a^2}{2} \frac{\alpha_\parallel}{K_\perp} Q \tag{3}$$

in which the parameter $\alpha_\parallel / K_\perp$ must be minimized. Note that the expression is independent of both the initial radius of curvaturn R and the mirror thickness. The latter conclusion is explained as follows: a given flux Q produces a temperature gradient ΔT proportional to the mirror thickness h. Through a cancelling effect the resulting mirror curvature is independent of h and the central deflection thus is the same, whatever the mirror thickness.

Change to Operating Temperature. From the manufacturing conditions at ambient temperature to operational state in orbit, the mirror experiences a drastic uniform change (usually drop) in temperature. An expression, similar to the one derived for uniform rise is given in terms of the temperature change ΔT_u to the operating temperature ($\sim T_o$). Since the coefficient of thermal expansion varies with T, an equivalent $\overline{\alpha}(T)$ appears in the equation:

$$\delta \cong \frac{a^2}{2R} \overline{\alpha}_\parallel (T) \Delta T_u \tag{4}$$

This deflection will take place only in the presence of an initial curvature $1/R$.

Variation in α. The inhomogeneity expected in the coefficient $\overline{\alpha}$ of most materials is of great concern especially because of its unknown distribution. Consider a simple model consisting of two equally thick layers of material with dissimilar α's. When exposed to a uniform temperature rise ΔT_u a flat mirror experiences a deflection proportional to the difference $\Delta\alpha$ between the respective coefficients α_1 and α_2 of the two layers.

$$\delta = \frac{3}{4} \frac{a^2}{m} \rho \Delta\alpha_\parallel \Delta T_u \tag{5}$$

Out-of-Plane Growth. A uniform temperature change ΔT_u causes the mirror to grow uniformly in thickness ($h = m/\rho$) by the amount

$$\delta = m \rho^{-1} \overline{\alpha} (T) \Delta T_u \tag{6}$$

Positioning actuators may be needed to compensate for this deflection by a rigid body movement.

Dynamic Loading

The resonant peak deflection of the mirror excited by a harmonic disturbance at its center is suggested as a basis of comparison. A damping ratio $\xi = .05$ is selected constant for all materials for lack of better information at the moment.

The disturbing function is a sinusoidal loading

$$F(t) = \overline{F} \sin \omega_1 t$$

where $\overline{F}$ is the amplitude of the disturbance and ω_1 is equal to the fundamental frequency of the mirror considered. Noting that the fundamental frequency of the circular plate is

$$\omega_1 = 1.52 \; \frac{m}{a^2} \sqrt{\frac{E}{\rho^3(1-\nu^2)} \; \frac{I}{I_s}} \; ,$$

it follows that the resonant peak mirror deflection, including the contribution of the first mode only (conservative) is

$$\delta = .22 \; \frac{a^2}{m^3} \; \frac{\rho^3 (1-\nu^2)}{E_\parallel} \left(\frac{I_s}{I}\right) \frac{\overline{F}}{\xi} \tag{7}$$

Thus the material parameter affecting the dynamic response is the same as the one appearing in the static case (Equation 1).

Transient Thermal Effects

For the case of transient response to the sudden application of a thermal flux, the sag variation as a function of time t becomes important. Assuming a constant absorbed flux Q suddenly applied to the front face of the mirror having an ideally insulated back face, the average rate of sag, between time $t = 0$ and the time t_1 at which the back face starts rising in temperature, is given by

$$\frac{\delta}{t} \cong \frac{2a^2}{m^2} \; \frac{\rho \, \alpha_\parallel}{C_p} \; Q \tag{8}$$

example, δ_2 and δ_4 (important in Case A) can be minimized by using nearly flat mirrors (R $\longrightarrow \infty$).

8. The deflection due to a linear gradient δ_3 is significant in Case B, and essentially removes materials such as LI900, and glassy C from prime consideration here. If the negative sign of δ_3 is taken into account, $-\delta_3$ for SiO_2 cancels most of the other deflections to bring it to second place behind C/gr.

9. An analogous consideration to item 2 suggests that if δ_6 can be corrected for, then graphite with a low modulus metal matrix is the most promising material for high flux situations. On the basis of δ_1, δ_2, δ_3, δ_4, and δ_7 only, Gr/Mg-14% Li shows only a total absolute deflection of 0.51 microns, compared to C/gr (1.33), Be (1.6) and Gr/epoxy (1.8) (the next best materials).

10. While repositioning mirrors under high flux conditions may be difficult (to overcome δ_6 deflections), one notes, analogous to item 3, that larger high flux mirrors will have relatively smaller deflections due to δ_6. Gr/Mg-Li and Be then appear particularly promising.

11. Composites could usefully employ graphite fibers perpendicular to the mirror plane, both for lowering α_1 and raising K_1. Carbon graphite represents an approach to this ideal.

12. Al, SiC, Gr/Al, Be and Invar are prime candidates when speed of response to a given flux is important in a mirror (Table III).

DISCUSSION

The assumptions used for calculating δ_2 and δ_3 are justified in all cases except one. The temperature gradient $(T_f - T_b)$ equals $Qm/\rho K_1$ and is <2°K for all situations except LI900 when $Q \sim 400w/in^2$. Here the high h/K_1 ratio requires a high T_f value and re-radiation is expected. In this case we estimate δ_3 by equating the non-reflected incident radiation with the sum of the re-radiated and the conducted heat flux under steady state conditions

$$Q = \sigma \epsilon \, T_f^4 + K \, \rho \, (T_f - T_b) \, m^{-1} \qquad (9)$$

Allowing a 50% increase in K due to the higher average temperature one obtains $T_f \sim 800°K$. The adjusted value of δ_2 is low ($\sim .04\mu$), but δ_3 is reduced only to $\sim 187\mu$ by this analysis (Fig. 3).

As time t increases beyond $t_1 \cong .08 \, C_p m^2/\rho K$ the mirror
merely experiences a uniform rise in temperature and the initial
gradient remains practically constant. The parameter $\rho \alpha/C_p$
can serve as a measure of the quickness with which the structure
will react to the sudden flux.

Materials Properties

The equations presented above furnish the engineer with a
useful tool to compare the respective merits of various materials.
It is apparent that materials properties such as α, $\Delta\alpha$, K, E and
ρ determine to a great extent the mirror deflections likely to be
encountered. Table I lists twelve candidate materials and their
relevant properties at the temperatures (T_o) under consideration.
Some of these materials are well known, others are developmen-
tal. Pseudo-isotropic graphite epoxy here represents an 8 unidirectional ply lay up, $(0/45/90/-45)_s$ of the Celanese GY70
fiber in Fiberite X-30 resin. Carbon graphite represents a wide
range of graphite fiber reinforced carbon or graphite composites.
The fibers are usually in some three dimensional arrangement.
LI900 is a low density high purity silica insulation material
developed by Lockheed for the space shuttle thermal protection
system. Graphite fibers in magnesium with 14% lithium (Ref. 1)
represents a hypothetical material with fifty volume percent
GY70 fibers. Optimum reported materials properties were used
whenever possible. The $\Delta\alpha$ values were estimated as a percentage of the α values. 2 ppb was the smallest $\Delta\alpha$ value considered
likely and was assigned to fused silica due to its compositional
uniformity. A wider range of materials is analyzed in Ref. 2.

RESULTS OF DEFLECTION CALCULATIONS

Non-materials parameters studied are listed in Table II.
Figures 2 and 3 summarize the results of calculations for cases
A and B respectively. Both the total and relative values of
deflections δ_1 through δ_7 are shown. Since $\Delta\alpha$ values are least well
known, and δ_5 values are relatively large, these deflections are
presented separately on the negative vertical axis. Significant
results are as follows:

1. In Case A, it is noted that if the negative values of δ_2 and
    δ_3 are considered (and the negative α_1 for ULE (Table I),
    then ULE gives a slightly lower total of deflections than SiO_2.

2. The out-of-plane growth (δ_6) is a major deflection for pseudo-
    isotropic composites. Since it can be corrected for by
    positioning actuators after T_o has been reached, especially

TABLE I. MATERIAL PROPERTIES

( ) Denote Estimates

Property	T_o ^oK	$\alpha_{\parallel}$ $^oK^{-1}$ 10^{-6}	$\Delta\alpha_{\parallel}$ $^oK^{-1}$ 10^{-8}	$\bar{\alpha}_{\parallel}$ $^oK^{-1}$ 10^{-6}	$\bar{\alpha}_{\perp}$ $^oK^{-1}$ 10^{-6}	$K_{\perp}$ $w\text{-}m^{-1}\text{-}{}^oK^{-1}$	C_p $J\text{-}Kg^{-1}\text{-}{}^oK^{-1}$ 10^2	ρ $Kg\text{-}m^{-3}$ 10^3	$E_{\parallel}$ $N\text{-}m^{-2}$ 10^{10}	ν
Factor	1					1				1
Al	150	17	12	20	20	248	6.4	2.7	7.5	.33
	350	24	"	23.8	23.8	237	9.0	"	7.1	"
Be	150	3.7	6	8	8	451	6.27	1.83	29.3	.05
	350	13	"	13	13	200	18.8	"	28.5	"
C/Gr	150	0.3	3	(0.3)	0.3	(120)	(6.9)	1.85	(10)	.04
	350	0.3	"	"	"	173	8.33	"	10.3	"
Glassy C	150	(2.2)	2	(3)	(3)	(7)	7.0	1.5	3.5	.2
	350	(3.2)	"	2.1	2.1	13	7.53	"	"	"
Gr/Al (V_f=.3)	150	(5)	5	(5)	(16)	(125)	6.2	2.39	(16)	.25
	350	5	"	"	(18.6)	"	7.5	"	"	"
Gr/ep (pseudo-iso.)	150	.07	0.7	.07	(40)	(0.7)	(8)	1.85	10.3	0.3
	350	"	"	"	39.6	1.4	8.4	"	"	"
Gr/Mg14Li (V_f=.5)	150	1.34	5	(1.1)	(19)	(60)	8.1	1.65	32	.3
	350	0.95	"	.95	(24)	(54)	10.3	"	31.5	"
Invar (LR 35)	150	1.4	0.6	(1)	(1)	(12)	(4)	8.1	15.5	.29
	350	0.6	"	0.6	0.6	12.6	5.15	"	14.8	"
Porous SiO$_2$ (LI 900)	150	-0.07	5	0.3	0.3	0.01	3.82	0.144	0.014	(.2)
	350	0.6	"	0.6	.6	0.015	6.28	"	0.02	"
SiC	150	(3)	4	(3.3)	(3.3)	140	2.55	3.2	(43)	(.2)
	350	3.6	"	3.6	3.6	71.2	7.61	"	42.8	"
SiO$_2$ (7940)	150	0.07	0.2	.21	.21	1.0	4.18	2.2	6.6	.14
	350	0.56	11	.56	.56	1.4	7.5	"	"	"
ULE (7971)	150	-.5	0.4	-0.26	-.26	(1)	(4.2)	2.2	(6.6)	(.17)
	350	0.03	"	0.03	0.03	1.31	7.7	"	"	"

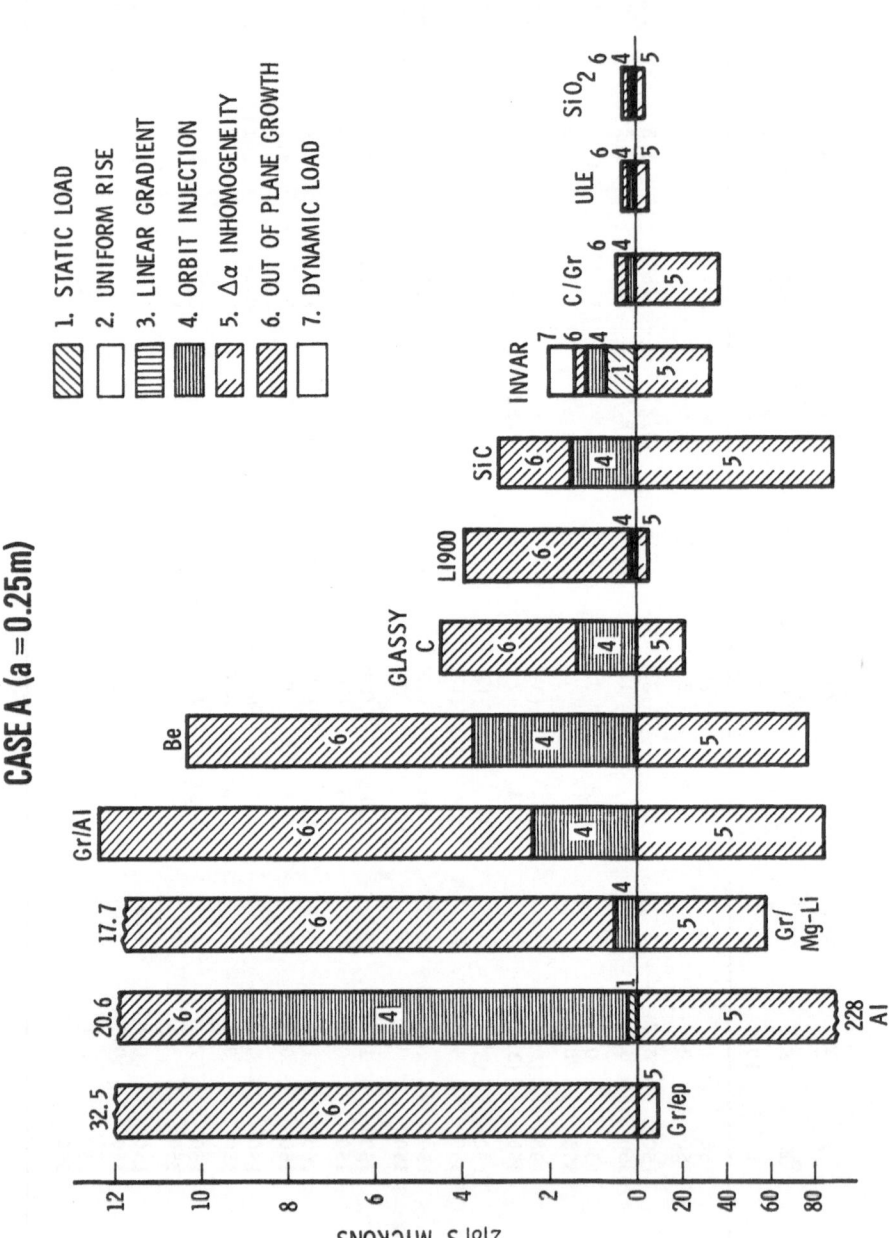

Figure 2.   Deflections of Various Materials in Low Radiation Flux Optics.

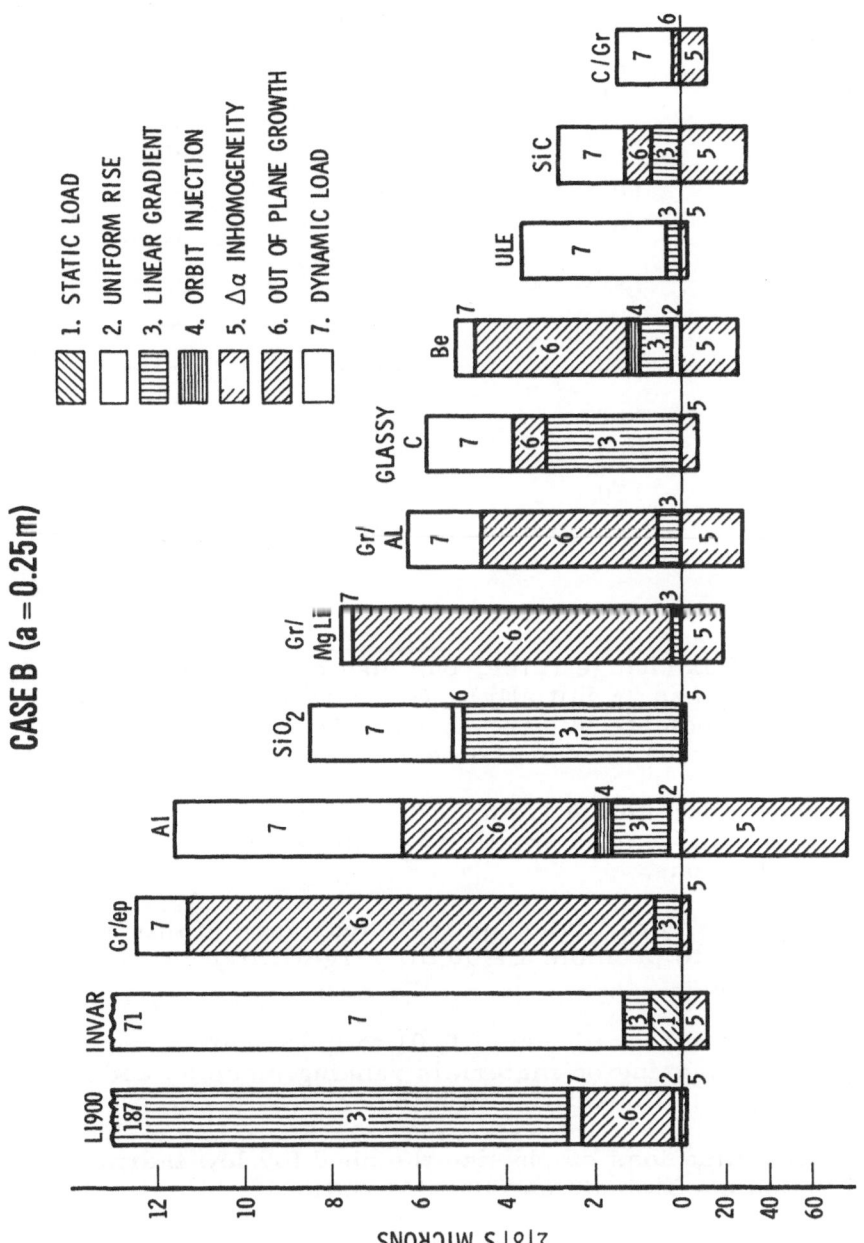

Figure 3. Deflections of Various Materials in High Radiation Flux Optics.

TABLE II

GEOMETRICAL AND LOADING CONDITIONS

Design Parameter	Case A	Case B	Units
a	0.25	0.25	m
R	10	100	m
m	10	10	$Kg-m^{-2}$
I/I_s	1300	1300	----
F_ϵ	1	1	----
ξ	0.05	0.05	----
$\overline{F}$	1	100	N
Q	0.01	400	$w-m^{-2}$
T_o	150	350	°K
ΔT_u	150	50	°K

for small flux change rates, one may compare materials by ignoring δ_6 (and δ_5 initially). Graphite-epoxy is then the foremost candidate in Case A, with C/gr, LI900, SiO_2 and ULE comparable second choices.

3. One should note that δ_6 is not size dependent so that the larger the mirror the smaller δ_6 will become relative to the total deflections.

4. The low density of LI900 and the importance of this property for δ_1 helps to promote LI900 for larger mirrors in Case A type conditions.

5. A change of absorbed flux of 0.01 to 0.15 $w-m^{-2}$ has no significant bearing on materials ranking since δ_2 and δ_3 are relatively small quantities in either case. (Case A).

6. Low flux situations emphasize the need for low thermal expansion coefficients and their variations.

7. The equations presented provide guidelines to the minimization of dimensional instabilities through design modifications as well as materials optimization and development. For

TABLE III

VALUES OF δ_8/t FOR VARIOUS MATERIALS

Case A		Case B	
Units of 10^{-11}m/s		Units of 10^{-6}m/s	
Al	(89.6)	Al	(36)
SiC	(47)	Gr/Al	(8)
Invar	(35.4)	SiC	(7.6)
Gr/Al	(24)	Be	(6.3)
Be	(13.4)	Invar	(4.7)
Glassy C	(5.9)	Glassy C	(3.2)
Gr/Mg-Li	(3.4)	SiO_2	(0.8)
ULE	(3.3)	Gr/Mg-Li	(0.76)
C/Gr	(1)	C/Gr	(0.33)
SiO_2	(0.5)	Gr/ep	(0.077)
Gr/ep	(0.2)	Li900	(0.055)
LI900	(0.03)	ULE	(0.043)

It is evident that the use of an I beam cross section would reduce deflections due to static and dynamic loads to the same order of magnitude as most thermally induced values. The limits to the increase of the ratio I/I_s are also a function of materials properties, principally the modulus of elasticity E and the yield strength F_y. Further analysis is needed to determine whether thin plates associated with a sandwich construction are subject to buckling instabilities.

Launch, deployment, or operating (slew) loads may also cause permanent deformations. For example, the maximum stress (σ_{max}) for the simply supported circular flat plate of diameter 2a and effective gravity g' is

$$\sigma_{max} = \frac{3(3+\nu)}{8m} \, a^2 \rho^2 \, g' \tag{10}$$

Insertion of typical values during launch suggests σ_{max} is not likely to exceed $\sim 3 \times 10^7 N/m^2$ (4350 psi). This is a modest requirement for most materials, especially graphite-epoxy that has not received much thermal cycling. However, more attention should be paid to σ - ϵ_p curves for $\epsilon_p < 10^{-6}$, for smaller deformations may be signiffcant when local tolerances on the order of $\lambda/100$ are required. For example, a stress of 4300 psi is sufficient to cause a permanent strain of 3×10^{-7} in Invar sheet (Ref. 3). Few data of this type are available for other materials.

With care and development, the absolute average values of α can be made to approach zero for ULE, Invar, and some composites, especially for $350^{\circ}K$ operation. However α variations within a structure may be much more important. Little work has been done to date on such measurement. Hagy and Shirkey (Ref. 4) show that (with ULE) α can be measured to an accuracy of 1-2 ppb/$^{\circ}K$ with a method (ultrasonic) that can also test local α variations. The ultrasonic method can be used to measure the average α in any direction. It has been used to measure vertical variations in α as well as those perpendicular to the top surface.

The choice of a first surface mirror or substrate material must also consider fabricability e.g. replication, availability, polishability, coatings adhesion/compatibility and optical requirements, degassing/moisture/contamination, joints/attachments/end effects, radiation/electrical/thermal environment, thermal cycling/residual stress/temporal stability characterization and cost.

For large mirrors with finite values of thermal expansion and elastic modulus, unacceptably large deflections may be produced by operational thermal and inertial environments. One remedy that has received considerable attention by NASA and DARPA is the use of active control of mirror figure and tilt (Ref. 5). In this technique servo-actuators either move the elements of a segmented mirror or apply local deformations to a thin mirror facesheet to compensate for random environmental distortions. Active control can be designed to minimize certain vibrational modes or total rms mirror deflection. By utilizing numerous actuators in both tilt and focus it is now possible to construct large ground- or space-based telescopes capable of producing images of nearly diffraction-limited quality in complex operational environments. However the development of low-expansion, high-stiffness optical materials can greatly lighten the load on the system designer and significantly contribute to overall system realiability.

ACKNOWLEDGEMENTS

The authors wish to thank W. J. Cuneo, Jr., E. G. Kendall, J. Lloyd, W. A. Plummer, R. L. Pleasant and W. C. Riley for helpful comments and suggestions.

REFERENCES

1. R. A. Munroe, "Magnesium-Lithium Alloy Lightens Electronic Packaging, Metal Progress 90 (1) p 89-92, July 1966.

2. E. G. Wolff, "Effect of Materials Properties on the Dimensional Stability of Large Space Optics", SAMSO-TR-(1977) (in preparation).

3. E. G. Wolff, C. S. Susskind, and D. L. Dull, "Effects of Thermal Mechanical Processing on the Microyield Strength of Invar", SAMSO TR-76-191, August 1976.

4. H. E. Hagy and W. D. Shirkey, "Determining the Absolute Thermal Expansion of Titanium-silica Glasses: a Refined Ultrasonic Method." Applied Optics 14 (9) 2099-2103 (1975).

5. W. E. Howell, "Recent Advances in Optical Control for Large Space Telescopes", in Space Optics, B. J. Thompson and R. R. Shannon Eds., National Academy of Sciences, Washington, D.C. 1974.

THERMAL EXPANSION AND CHEMICAL BOND

K.V. Krishna Rao and V.T. Deshpande

Department of Physics, Osmania University

Hyderabad - 500007, India

INTRODUCTION

It is the aim of the crystal physicist to explain the physical
properties of a crystal in terms of its structure and the bonding
between the constituent atoms and correlate the various physical
properties. In this laboratory, a number of physical properties
like elasticity, photoelasticity, and thermal expansion have been
extensively studied for the past twenty-five years with this aim
in view. It is felt that such a correlation is possible only if
the property is studied for a number of crystals belonging to
different crystal and bond types. A perusal of the literature[1,2]
showed that, in the case of thermal expansion, the data on a
family of crystals belonging to the same structure type were meagre.
Hence the authors have taken up the study of the thermal expansion
of a number of crystals belonging to a few important structure
types scheelite, ADP, rutile and calcite types. The X-ray method
has been used in this investigation in view of the advantages of
the method over other methods especially in the case of non-cubic
crystals. This paper presents a review of the results obtained.

EXPERIMENTAL

A Unicam 19 cm high temperature powder camera and a symmetrical back reflection camera constructed in the laboratory are used
to take the powder photographs of the substances in the temperature range 30°C to 700°C. Every attempt is made to record high
angle reflections for the precision determination of the lattice
parameters. In each case, about ten diffraction lines with Bragg

angles ranging from 55 to 80°C are used to evaluate the cell constants applying Cohen's[3] analytical method. Copper and iron radiation depending on the substance are used to take the powder photographs. The full experimental details and the results obtained have already been presented in the theses[4-8] and papers [9-13] published from this laboratory.

RESULTS AND DISCUSSION

Scheelite Type Compounds

The scheelite type compounds that have been studied are $CaWO_4$, $SrWO_4$, $BaWO_4$, $PbWO_4$, $CaMoO_4$, $CdMoO_4$, $SrMoO_4$, $PbMoO_4$, $NaIO_4$ and KIO_4. Table – I gives the data on the lattice parameters, the principal co-efficients of thermal expansion, the axial ratios and the anisotropy coefficients for these compounds at the lowest and at the highest temperatures of the range covered.

TABLE I

DATA ON SCHEELITE TYPE CRYSTALS

Compound	Temperature °C	a ($\overset{\circ}{A}$)	c ($\overset{\circ}{A}$)	$\alpha_{\perp} \times 10^6$	$\alpha_{\parallel} \times 10^6$	c/a	$\alpha_{\parallel}/\alpha_{\perp}$
$CaWO_4$	30	5.2437	11.3754	6.35	12.38	2.1639	1.950
	347	5.2587	11.4336	13.20	21.93	2.1742	1.661
$SrWO_4$	30	5.4189	11.9552	5.86	13.21	2.2062	2.254
	350	5.4319	12.0143	12.27	21.32	2.2118	1.738
$BaWO_4$	30	5.6138	12.7152	4.43	18.35	2.2649	4.142
	350	5.6234	12.8144	6.95	31.31	2.2788	4.505
$PbWO_4$	30	5.4691	12.0495	8.13	19.73	2.2056	2.427
	350	5.4773	12.1328	8.28	24.45	2.2151	2.953
$CaMoO_4$	30	5.2266	11.4343	7.67	11.88	2.1877	1.549
	350	5.2430	11.4941	12.60	21.20	2.1923	1.682
$CdMoO_4$	30	5.1560	11.1953	6.85	15.13	2.1713	2.209
	350	5.1724	11.2537	12.66	18.70	2.1757	1.477
$SrMoO_4$	30	5.3963	12.0234	6.90	17.03	2.2281	2.468
	650	5.4296	12.1863	12.96	26.15	2.2444	2.018
$PbMoO_4$	30	5.4362	12.118	8.73	20.35	2.2280	2.332
	350	5.4574	12.2065	15.93	26.89	2.2367	1.688

TABLE I (Continued)

Compound	Temperature °C	a (Å)	c (Å)	$\alpha_\perp \times 10^6$	$\alpha_{||} \times 10^6$	c/a	$\alpha_{||}/\alpha_\perp$
$NaIO_4$	30	5.3396	11.9559	38.17	54.87	2.2391	1.438
	120	5.3606	12.0273	45.07	71.33	2.2436	1.583
KIO_4	30	5.7315	12.6050	25.35	56.62	2.1922	2.234
	80	5.7441	12.6594	65.27	128.42	2.2039	1.968

It can be seen from Table I that all the crystals exhibit the same type of anisotropy in the coefficients of thermal expansion. In every case, the coefficient of thermal expansion along the tetragonal axis, $\alpha_{||}$ is larger than the coefficient along the perpendicular direction, $\alpha_\perp$. The same behaviour persists at all temperatures. The substances behave like layer structures with the layers normal to the tetragonal axis. The coefficients of expansion for the monovalent periodates are higher than those for the divalent crystals as is to be expected. In general, the anisotropy coefficient $\alpha_{||}/\alpha_\perp$ is found to increase with increasing cationic size.

In what follows, the thermal behaviour of these compounds is explained on the basis of the structure and the disposition of the bonds. The scheelite structure belongs to the space group $14_1/a$ and contains four formula units (ABO_4, A = Ca, Sr, Ba, Pb, Cd, Na or K and B = W, Mo or I) per unit cell (Figure 1). The structure may be considered as a layer structure (Arbel and Stokes, 1964)[14] the layers being perpendicular to the tetragonal axis. The ions of either sign form a simple square structure in the layer. The two square lattices interpenetrate in such a way that the ions of the same sign are situated at the corners of the square formed by the ions of the opposite sign. The three dimensional crystal is formed by stacking such layers, one above the other with a relative displacement of a/2 in the <100> direction. The unit cell comprises of four such layers. Thus each ion in the crystal from a given layer is surrounded by eight ions of the opposite sign. Four of these are from the same layer and other four are from the two adjacent layers two from each and form a tetrahedral configuration about the central ion. The electrostatic attraction between the central ion and four of its immediate neighbours of opposite sign from the same layer will provide binding forces entirely lying in the layers. On the other hand, the immediate neighbours from adjacent layers will interact with the central ion with forces

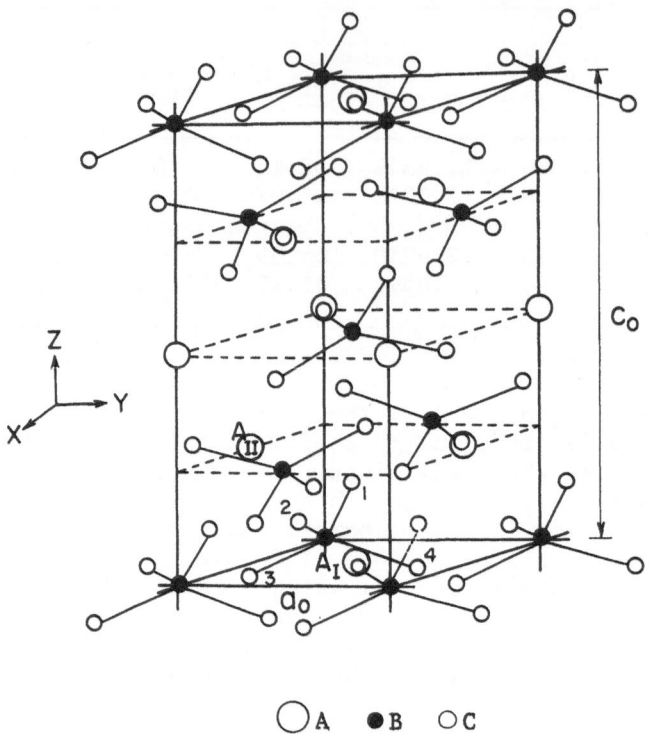

$$\bigcirc A \quad \bullet B \quad \bigcirc C$$

Fig. 1 The Structure of ABC_4 Compounds of Scheelite Type

which will have components both in the layers and in the perpendicular direction. Thus the layers can be said to be more strongly
bound than the inter-layer binding. This explains the characteristic thermal behaviour of the scheelite type compounds.

In these crystals, there is a close correlation between the
elastic constants and the coefficients of thermal expansion. For
instance, the linear compressibilities[15] for $CaWO_4$ along the
tetragonal axis and in the perpendicular direction $K_{||}$ and $K_{\perp}$
respectively, (Wachtman et al., 1968[15]) are

$$K_{||} = 4.98 \times 10^{-13} \text{ cm}^2/\text{dyne}$$

$$K_{\perp} = 3.64 \times 10^{-13} \text{ cm}^2/\text{dyne}$$

showing that the strength of the binding along the unique axis is less than that in the perpendicular direction. The elastic anisotropy ($K_{||}/K_{\perp}$ = 1.368) is of the same order as that of the thermal expansion ($\alpha_{||}/\alpha_{\perp}$ = 1.549).

KDP Type Crystals

The KDP type crystals that have been studied are potassium dihydrogen phosphate (KDP), ammonium dihydrogen phosphate (ADP) rubidium dihydrogen phosphate (RDP), ammonium dihydrogen arsenate (ADA) and potassium dihydrogen arsenate (KDA). Table - II gives the values of the lattice parameters at room temperature, the axial ratios, the values of the elastic stiffness constants c_{11} and c_{33} and the coefficients of thermal expansion. The values of c_{11} and c_{33} are taken from Haussühl (1964)[16]. $\overline{\alpha}_{\perp}$, $\overline{\alpha}_{||}$, and $\overline{\alpha}_{v}$ are the mean coefficients of expansion in the range of temperatures investigated.

TABLE II

DATA ON KDP TYPE CRYSTALS

Crystal	a (Å)	c (Å)	c/a	$c_{11} \times 10^{-11}$ dyne cm^{-2}	$c_{33} \times 10^{-11}$ dyne cm^{-2}	$\overline{\alpha}_{\perp} \times 10^{6}$	$\overline{\alpha}_{||} \times 10^{6}$	$\overline{\alpha}_{v} \times 10^{6}$
KDP	7.452	6.9683	0.9350	7.165	5.640	26.9	46.6	100.4
ADP	7.5006	7.5494	1.0065	6.877	3.402	39.3	1.9	80.5
RDP	7.6063	7.3009	0.9599	6.697	5.296	27.8	52.0	107.6
ADA	7.7003	7.7158	1.0020	6.220	2.956	22.3	1.0	45.6
KDA	7.6300	7.1630	0.9388	6.482	4.824	24.9	50.0	99.8

Table - II shows that, in the two crystals containing the ammonium radical, the axial ratio c/a is more than one whereas for the other three crystals, the ratio is less than one. The elastic constant c_{33} is smaller than c_{11} in all the crystals but the disparity between c_{11} and c_{33} is much greater in the crystals containing the ammonium radical the values of c_{11} being nearly double the values of c_{33}. In the case of the potassium and rubidium salts, $\alpha_{||}$ is greater than $\alpha_{\perp}$, whereas for the ammonium salts, $\alpha_{\perp}$

is many times larger than $\alpha_{||}$. Hence, it is obvious that the two
ammonium salts have some properties which are very much different
from the other members of the isomorphous series.

    This peculiarity in the behaviour of ADP and ADA may be
attributed to the presence of NH_4^+ ions and their ability to form
hydrogen bonds. In order to understand the special role of the
ammonium ions and to explain the anomalous behaviour of the
ammonium salts, it is necessary to look into the structural details
of the KDP type crystals (Fig. 2) The OX_4 groups (X = P or As)
are situated at the corners, at the centre of the unit cell and
also on the four faces parallel to the c-axis at (1/2, 0, 1/4),
(1/2, 1, 1/4), (0, 1/2, 3/4) and 1, 1/2, 3/4). The cations are
situated at distances c/2 apart from the XO_4 groups along the c-axis.
Each anionic group is linked to four similar neighbouring groups
by O-H...O bonds. Each cation in the lattice is surrounded by
eight oxygen atoms which form two interpenetrating tetrahedra. One
of the tetrahedra is flat while the other is steep with respect to
the c-axis. In ADP and ADA, the oxygen atoms of the flat tetra-
hedron are linked to the nitrogen atom of the NH_4^+ ion by N-H....O
bonds. In all the crystals, the O-H....O bonds linking XO_4 groups
are only slightly inclined to the XY plane. This explains why $\alpha_\perp$
is smaller than $\alpha_{||}$ in KDP, RDP and KDA. In the case of ADP and ADA,
the extra hydrogen bonds i.e. N-H.....O also are only slightly inclined to the XY plane, so that the strength of binding in XY plane
increases still further.

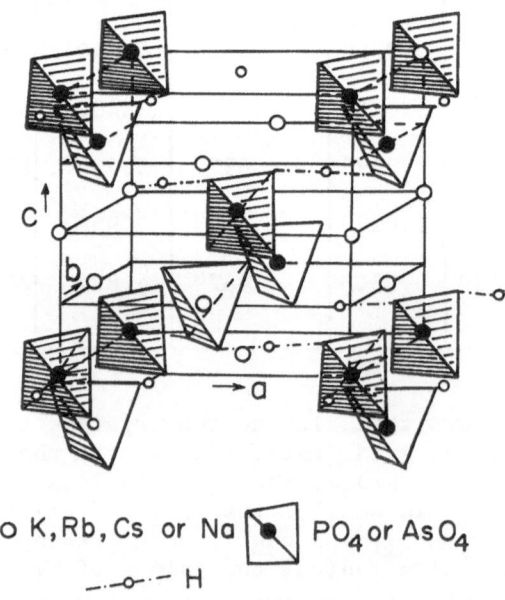

o K,Rb,Cs or Na PO_4 or $As O_4$
    ---o--- H

Fig. 2   Structure of KDP-Type Crystals

The additional strength given to these crystals in the XY plane is likely to produce a contraction of the lattice in this plane and a corresponding Poisson type of elongation along the c-direction. This explains the axial ratio in the case of ADP and ADA being greater than one. This is also consistent with the relative values of the elastic stiffness constants c_{11} and c_{33} and the smaller values of volume coefficient of ADP and ADA. However, the directional thermal expansion behaviour is not consistent with this picture. Due to the presence of the additional N-H.....O bonds one would expect that $\alpha_\perp$ would be much smaller than $\alpha_{||}$ in ADP and ADA. But it is actually found that $\alpha_\perp$ is many times larger than $\alpha_{||}$ in these crystals. In what follows, an attempt is made to explain this abnormal behaviour.

The explanation is based on the suggestion made by Tenzer et al (1958)[17] that, in the case of the compounds containing the ammonium ion, the structure is under some kind of constraint due to the size and the shape of the ammonium ion. The abnormal feature due to the presence of the ammonium ion is the angle between XO and O-H....O bonds, i.e. the angle $X-O_H$....O has a higher value in the case of ammonium salts. For instance, the angle $P-O_H$.....O has a value of 116° 42' in ADP and 113° 15' in KDP. As the temperature increases, there is an expansion of the lattice accompanied by a general loosening of the structure, so that the constraints are reduced. As a consequence, the angle $X-O_H$....O tends to decrease towards the value obtaining in other crystals. The effect of the decrease in the angle $X-O_H$.....O, is to bring about a reduction in the value of the c-parameter. This reduction in the value of c superimposed on the usual thermal expansion would effectively reduce the observed coefficient of expansion in this direction.

Rutile Type Crystals

The results on the rutile and the calcite type compounds have been communicated to the 1973 International Thermal Expansion Symposium[18]. A brief summary of the results and an account of the more recent work are given here. The rutile type compounds that have been studied are MgF_2, MnF_2, ZnF_2, CoF_2, NiF_2, FeF_2, RuO_2, CrO_2, TiO_2, SnO_2, PbO_2, VO_2, GeO_2 and recently OsO_2. Table III shows $\alpha_{||}$, $\alpha_\perp$, the coefficient of cubical expansion evaluated from $\alpha_{||}$ and $\alpha_\perp$, and c/a for all the crystals at the room temperature and at the highest temperature studied.

TABLE III

DATA ON RUTILE TYPE CRYSTALS

Crystals	Temperature °C	$\alpha_{||} \times 10^6$	$\alpha_{\perp} \times 10^6$	$\Delta \times 10^6$	c/a
MgF_2	25	13.6	9.4	32.4	0.6607
	610	21.9	17.5	56.9	0.6624
MnF_2	25	12.9	3.6	20.1	0.6793
	450	18.9	7.7	34.3	0.6826
ZnF_2	25	11.6	8.7	29.0	0.6661
	450	19.0	11.3	41.6	0.6680
CoF_2	25	3.9	10.3	24.5	0.6767
	530	12.4	15.8	44.0	0.6750
NiF_2	25	9.0	7.2	23.4	0.6632
	570	14.8	15.3	45.4	0.6633
FeF_2	25	- 0.1	16.6	33.1	0.7050
	410	-10.2	28.3	46.4	0.6963
RuO_2	25	- 1.4	6.9	12.4	0.6917
	650	- 5.7	14.4	23.1	0.6849
CrO_2	25	-15.0	18.7	22.4	0.6597
	370	- 0.1	13.5	26.9	0.6543
TiO_2	25	8.9	7.3	23.5	0.6551
	610	13.4	9.1	31.6	0.6451
SnO_2	25	3.7	3.3	10.3	0.6725
	570	6.6	5.8	18.2	0.6728
PbO_2	25	9.3	8.4	26.1	0.6832
VO_2	80	27.5	5.3	38.1	0.6260
	380	22.2	4.1	30.4	0.6301
GeO_2	25	1.7	6.1	13.9	0.6509
	650	5.6	11.6	28.8	0.6489
OsO_2	25	0.0	2.1	4.2	0.7071
	410	0.0	19.4	38.9	0.7051

Though all the crystals belong to the same structure type, the thermal expansion shows large variations both in the values of the coefficients of thermal expansion and in the anisotropy of the coefficients. These variations in the coefficients of thermal expansion could be explained to a large extent in terms of the percentage ionic character and the inter ionic distances[19].

The percentage ionic character of the bond between two atoms A and B is calculated using the following equation given by Pauling (20).

Amount of ionic character = $1 - e^{-\frac{1}{4}(X_A - X_B)^2}$, where X_A and X_B are the electronegativities of the atoms A and B respectively. It has been found that broadly, the coefficients of thermal expansion are found to increase with the percentage ionic character.

Figure 3 shows the unit cell in rutile structure and the bonds A_1, A_2, S_1 and S_2 which bind the ions in the crystal. The strength of a bond is qualitatively estimated by comparing the bond length with the sum of the ionic radii. If the bond length is smaller than the sum of the ionic radii, it is concluded that the bond is strong. On the other hand, if the bond length is equal to or greater than the ionic radii the bond is considered as weak. Using this criterion, it has been possible to explain the anisotropy in thermal expansion in most of the compounds for which the electronic configuration of the cation has spherical symmetry.

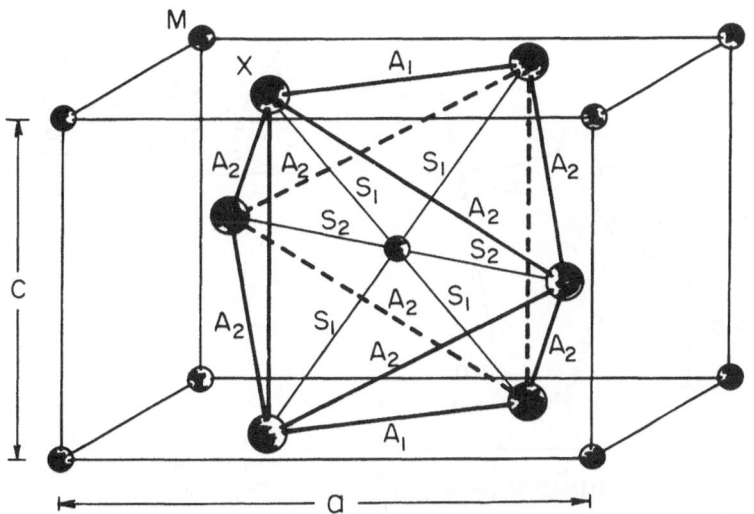

Fig. 3   Unit Cell of Rutile Type Structure
● Cation (M)   ◕ Anion (X)

It may be mentioned that the dependence of the thermal expansion behaviour on the electronic configuration of the ions has been pointed out for the first time by Leela Iyengar (1970)[6] in rutile type compounds. According to the electronic configuration of the d-shell of the cation, the compounds can be divided into three categories, the first category with spherical distribution, the second with octahedral distribution and the third with unsymmetrical distribution. The compounds in which the electronic configuration is unsymmetrical are found to show abnormal behaviour. A plot of the axial ratio versus number of d-electrons (Fig. 4) for each of the three transition metal dioxides and difluorides shows a peak corresponding to the maximum axial ratio for the series. The crystals at the maximum position are CrO_2, FeF_2, RuO_2 and OsO_2. It is interesting to note that earlier studies on the first three compounds showed that they have a negative coefficient of thermal expansion along the c-axis. Recent studies on OsO_2 showed that the crystal has zero coefficient of expansion along the c-axis. Since the axial ratio is high for these compounds, the c/a value decreases with increase of temperature by having a negative or zero coefficient of thermal expansion along the c-axis probably to acquire the stability corresponding to the normal c/a value. Hazony and Perkins (1970)[21] have explained the anomalous thermal expansion in FeF_2 and VO_2 in terms of the electronic structure of the cation.

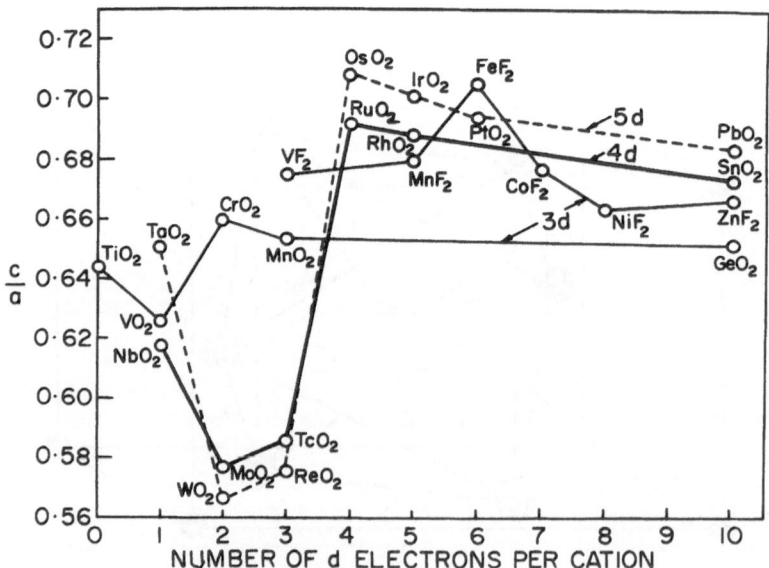

Fig. 4   c/a Versus Number of d Electrons per Cation

Calcite Type Crystals

The calcite type crystals that have been studied are $CaCO_2$, $CdCO_3$, $MnCO_3$, $ZnCO_3$, $NaNO_3$, paraelectric phase of KNO_3 and $ScBO_3$. The coefficients of thermal expansion, the interionic distances and the inclination of the M-O bond to the c-axis are given in Table IV.

TABLE IV

DATA ON CALCITE TYPE CRYSTALS

Crystal	Temperature °C	$\alpha_{||} \times 10^6$	$\alpha_{\perp} \times 10^6$	X-O(Å)	M-O(Å)	Inclination of M-O bond to the c-axis	
$CaCO_3$	25	25.10	-3.68	1.283	2.303	55°	0'
$CdCO_3$	25	18.98	-2.13	1.231	2.265	54°	55'
$MnCO_3$	25	21.76	1.34	1.288	2.140	54°	13'
$ZnCO_3$	25	23.51	9.07	1.280	2.073	54°	17'
$NaNO_3$	25	113.35	7.30	1.245	2.347	55°	1'
KNO_3	150	243.21	-8.52				
$ScBO_3$	25	9.72	1.24	1.350	2.089	55°	4'

Table IV shows that, in all the crystals, the coefficient of thermal expansion along the c-axis is larger than the coefficient in the perpendicular direction. In three crystals, $CaCO_3$, $CdCO_3$ and KNO_3, $\alpha_{\perp}$ is negative. The nitrates, $NaNO_3$ and KNO_3, the monovalent nitrates, have relatively large coefficients of expansion as is to be expected.

In the calcite structure MXO_3 (Figure 5) where M stands for the metal ion and X for carbon, nitrogen or boron, the planar XO_3 groups are perpendicular to the c-axis. It is usual, for a layer structure that the coefficient of expansion normal to the layers is large when compared to the coefficient of expansion along the layers. This broadly explains the thermal behaviour of these compounds. This is further elaborated in terms of the strength of the bonds as follows.

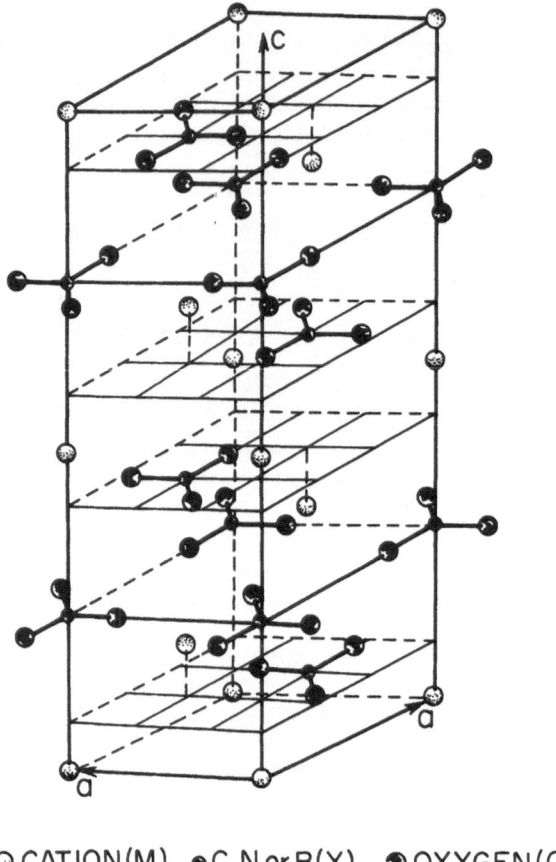

⊙ CATION(M)   ● C,N or B(X)   ◉ OXYGEN(O)

Figure 5 Unit Cell of Calcite Type Structure

Table IV shows that, for all the crystals, the lengths of the
observed X-O bonds are considerably less than those calculated on
the basis of the ionic radii. Hence the X-O bonds are tightly
bound within themselves by the formation of a planar triad of strong
co-valent bonds. On the other hand, the M-O bond distances are only
slightly less than those calculated on the basis of the ionic radii.
Hence the X-O bond is expected to be much stronger than the M-O
bond. The X-O bonds lie in planes normal to the c-axis and the M-O
bond is inclined at an angle of about 55° with respect to the c-axis.
Hence, on account of the strong X-O bonds in the basal planes and
the major contribution of the M-O bonds along these planes, the atoms
in the basal plane are more tightly bound than in the perpendicular

direction. This is in agreement with the observed anisotropy in thermal expansion.

CONCLUSION

In conclusion, it may be mentioned that there is close correlation between the strength of the chemical bond and the thermal expansion. In general, the coefficient of thermal expansion increases with the percentage ionic character of the bond. However, there are several other factors that contribute to the thermal expansion, for instance the release of the strains in the structure at elevated temperatures and the consequent change in the orientation of the bonds and groups of atoms. Some times the electronic configuration of the constituent ions may play an important role as in the case of the rutile type compounds.

ACKNOWLEDGEMENTS

The authors wish to thank their former studens, Drs. D.B. Sirdeshmukh, S.V.N. Naidu, A.A. Khan, S.V. Suryanarayana, K.S. Murthy and Leela Iyengar for their collaboration in this investigation.

REFERENCES

1. Wooster, W.A., (1949), "A Test Book on Crystal Physics", Cambridge University Press, Cambridge.

2. Srinivasan, R. and Krishnan, R.S., (1958), "Progress in Crystal Physics", S. Viswanathan, Madras, India.

3. Cohen, M.U. (1935), Rev. Sci. Instrum. $\underline{6}$, 68.

4. Khan, A.A. (1966), "X-ray Studies on Crystals", Ph.D. Thesis, Osmania University, Hyderabad, India.

5. Naidu, S.V.N., (1966), "X-ray Studies on Rutile Type Compounds", Ph.D. Thesis, Osmania University, Hyderabad, India.

6. Leela Iyengar, (1970), "X-ray Studies on the Thermal Expansion of Rutile Type and Related Compounds", Ph.D Thesis, Osmania University, Hyderabad, India.

7. Murthy, K.S.N. (1971), "X-ray Studies on the Thermal Expansion of Calcite Type Compounds", Ph.D. Thesis, Osmania University Hyderabad, India.

8. Suryanarayana, S.V., (1971), "X-ray Studies on some Scheelite
    Type Compounds", Ph.D. Thesis, Osmania University, Hyderabad,
    India.

9. Krishna Rao, K.V., Naidu, S.V.N., and Setty, P.L.N., (1962).
    Acta. Cryst. 15, 528.

10. Deshpande, V.T., and Khan, A.A., (1963), Acta. Cryst. 16 936.

11. Krishna Rao, K.V. and Leela Iyengar, (1969), Act. Cryst. A 25,
    302.

12. Deshpande, V.T. and Suryanarayana, S.V. (1972). Act. Cryst.
    A 23, 94.

13. Krishna Rao, K.V. and Murthy, K.S.N. (1970), J. Materials
    Science, 5, 82.

14. Arbel, A. and Stokes, R.J., Honeywell Research Report, (1964),
    H.R. 64.-375 Hopkins, U.S.A.

15. Wachtman, Jr. J.B., Brower, J.R.W.S., and Farabaugh, E.N.,
    (1968), J. Amer. Ceram. Soc. 51, 341.

16. Haussühl, V.S., (1964), Z. Krist. 120, 401.

17. Tenzer, L. Frazer, B.C. and Pepinsky, R. (1958), Acta. Cryst.
    11, 505.

18. Krishna Rao, K.V. (1974) AIP Conference Proceedings No. 17,
    American Institute of Physics, New York, pp. 219-230.

19. Krishna Rao, K.V. (1969), Physics of the Solid State Academic
    Press, pp. 415-425.

20. Pauling, L. (1960), "The Nature of the Chemical Bond", Cornell
    University Press, New York, p. 98.

21. Hazony, Y. and Perkins, H.K., (1970), J. Appl. Phys. 41, 5130.

THIRD-ORDER ELASTIC CONSTANTS AND THERMAL EXPANSION OF SCANDIUM

R.Ramji Rao and A.Ramanand

Physics Department

Indian Institute of Technology, Madras-600036,INDIA

ABSTRACT

The third-order elastic (T.O.E.) constants, the pressure derivatives of the second-order elastic (S.O.E.) constants and the temperature variation of the volume Gruneisen function of scandium have been calculated using the nearest-neighbour central force model for hcp metals proposed by Srinivasan and Ramji Rao. The T.O.E.constants have been employed to calculate the low-temperature limit $\bar{\gamma}_L$ of the lattice thermal expansion. The theoretical value of $\bar{\gamma}_L$ for scandium is 0.94. The high temperature limit of the lattice thermal expansion, $\bar{\gamma}_H$ has the value 1.22 which agrees well with the experimental value 1.17 obtained from the thermal expansion and lattice specific heat data of scandium. The variation of the lattice parameters of scandium with hydrostatic pressure has been calculated using its T.O.E. constant data and the extrapolation formula of Thurston. The agreement with the compression data of Stephens is good.

INTRODUCTION

The second-order elastic (S.O.E.) constants of scandium were measured by Fisher and Dever (1). The high temperature limit of the Gruneisen function, $\bar{\gamma}_H$ for scandium is 1.17 and has been calculated from thermal expansion and C_V data by Gschneidner (2). Stephens (3) has measured the compression of this metal upto 44 K-bars. In this paper the third-order elastic (T.O.E.) constants, the pressure derivatives of the

SOE constants and the temperature variation of the volume Gruneisen function $\bar{\gamma}_V^\zeta(T)$ of scandium have been calculated using the nearest-neighbour central force (C.F.) model of Srinivasan and Ramji Rao (4,5). Thurston's (6) extrapolation formula and the theoretical TOE constants have been used to calculate the changes in the lattice parameters "a" and "c" and in the volume "V" with hydrostatic pressure.

THIRD-ORDER ELASTIC CONSTANTS

The expressions for the TOE constants of an ideal hcp lattice were derived by Ramji Rao and Srinivasan (5) for the nearest-neighbour central interaction potential of the form given by

$$\Phi(r) = -(a/r^m) + (b/r^n) \qquad \qquad .. (1)$$

The TOE constants' expressions have been given in a consolidated form in the paper on Cobalt (7) and they involve two constants k_2 and k_3. We have

$$k_2 = 0'^2\Phi = \eta M/2D^2$$

and $$k_3 = 0'^3\Phi = -(\eta M/4D^4)(n + m + 6) \qquad .. (2)$$

where M is the mass of the atom, D is the lattice constant in the basal plane,

$$\eta = nb(n-m) / 2MD^{n+2} \qquad \qquad .. (3)$$

and $0'$ is the differential operator $\partial/\partial r^2$.

Srinivasan and Ramji Rao (4) showed that the high temperature limit of the Gruneisen function, $\bar{\gamma}_H$, of an ideal hcp lattice described by the potential $\Phi(r)$ defined in Eqn.(1) is nearly equal to $(n + m)/6$. The value of $(n + m)$ pertinent to scandium is obtained as 7.02, using Gscheidner's (2) reported value of 1.17 for $\bar{\gamma}_H$ of scandium. The S.O.E. constants C_{11} and C_{33} at 298°K measured by Fisher and Dever (1) and the lattice parameter D = 3.302 a.u. are used to get an average value of η for this metal. The theoretical TOE constants of scandium at 298 K are presented in Table 1.

The expressions for the effective SOE constants C'_{ij} of a strained hexagonal solid were derived by Ramji Rao and Srinivasan (8) in terms of its TOE constants using the finite strain theory of Murnaghan (9). The pressure derivatives of the SOE constants of scandium at 298 K are obtained by

Table 1:   TOE Constants and the Pressure Derivatives of SOE
Constants of Scandium at 298 K

C_{ijk}	Value in 10^{11} dyn/cm^2	Pressure derivatives of C_{ij}		
C_{111}	$-$ 89.6	$\partial C_{11}/\partial p$	$=$	4.60
C_{222}	$-$ 109.5	$\partial C_{12}/\partial p$	$=$	2.65
C_{333}	$-$ 94.0	$\partial C_{13}/\partial p$	$=$	1.95
C_{112}	$-$ 31.5			
C_{113}	$-$ 4.5	$\partial C_{33}/\partial p$	$=$	5.73
C_{123}	$-$ 7.2			
C_{133}	$-$ 23.5	$\partial C_{44}/\partial p$	$=$	0.96
C_{144}	$-$ 5.9			
C_{155}	$-$ 5.9	$\partial C_{66}/\partial p$	$=$	0.98
C_{344}	$-$ 23.5			

differentiating C'_{ij} with respect to pressure and using the
calculated TOE constants. The $\partial C_{ij}/\partial p$ values of scandium
are also given in Table 1.

THERMAL EXPANSION

Low Temperature Limit, $\bar{\gamma}_L$ of the Lattice Thermal Expansion

     The low temperature limit of the lattice volume
Grüneisen function, $\bar{\gamma}_L$ can be calculated from the generalized
Grüneisen Parameters (GPs) of the elastic waves propagating
in the crystal. The GPs of the normal mode frequencies of a
uniaxial crystal are given by the expressions

$$\gamma'(\omega) = -(1/\omega)(\partial\omega/\partial\epsilon'), \quad \gamma''(\omega) = -(1/\omega)(\partial\omega/\partial\epsilon'') \; . \; (4)$$

Here ϵ' is a uniform areal strain perpendicular to the unique
axis and ϵ'' is a uniform longitudinal strain parallel to the
hexagonal axis. We now define the effective Grüneisen func-
tions $\bar{\gamma}_\perp^t(\tau)$ and $\bar{\gamma}_\parallel^t(\tau)$ for a uniaxial crystal as the

weighted averages of the generalised GPs. Thus

$$\bar{\gamma}_{\perp}^{\ell}(T) = \left[\sum_{\vec{q},j} \gamma'(\vec{q},j) \, C_v(\vec{q},j) \right] \Big/ \sum_{\vec{q},j} C_v(\vec{q},j)$$

$$\bar{\gamma}_{\parallel}^{\ell}(T) = \left[\sum_{\vec{q},j} \gamma''(\vec{q},j) \, C_v(\vec{q},j) \right] \Big/ \sum_{\vec{q},j} C_v(\vec{q},j) \quad ..(5)$$

where $\vec{q}$ is the wave vector, j is the polarisation index and $C_v(\vec{q},j)$ is the contribution to the lattice specific heat of a single normal mode of frequency $\omega(\vec{q},j)$. The superscript "ℓ" on the gammas indicates that these are lattice Gruneisen functions. At very low temperatures the elastic modes predominate and $\bar{\gamma}_{\perp}^{\ell}(T)$ and $\bar{\gamma}_{\parallel}^{\ell}(T)$ approach the limits $\bar{\gamma}_{\perp}^{\ell}(-3)$ and $\bar{\gamma}_{\parallel}^{\ell}(-3)$ respectively, defined as

$$\bar{\gamma}_{\perp}^{\ell}(-3) = \left[\sum_{j=1}^{3} \int \gamma_j'(\theta,\phi) \, V_j^{-3}(\theta,\phi) \, d\Omega \right] \Big/ \sum_{j=1}^{3} \int V_j^{-3}(\theta,\phi) \, d\Omega$$

$$\bar{\gamma}_{\parallel}^{\ell}(-3) = \left[\sum_{j=1}^{3} \int \gamma_j''(\theta,\phi) \, V_j^{-3}(\theta,\phi) \, d\Omega \right] \Big/ \sum_{j=1}^{3} \int V_j^{-3}(\theta,\phi) \, d\Omega$$

$$\cdot\cdot(6)$$

Here $V_j(\theta,\phi)$ is the wave velocity of the elastic mode of polarisation index j propagating in the direction (θ,ϕ); $\gamma_j'(\theta,\phi)$ and $\gamma_j''(\theta,\phi)$ are the GPs for this elastic mode. The individual GPs of the elastic waves can be calculated from the TOE constants of the crystal using the procedure suggested by Ramji Rao and Srinivasan (5,10). In hexagonal crystals, the acoustic wave velocities and the GPs depend only on θ and are independent of the aximuth ϕ. Table 2 gives the wave velocities and the GPs for the acoustic waves propagating at different angles θ to the Z-axis in the X-Z plane. The calculated values of $\bar{\gamma}_{\perp}^{\ell}(-3)$ and $\bar{\gamma}_{\parallel}^{\ell}(-3)$ for scandium are 1.0 and 0.84 respectively.

The temperature variation of the linear thermal expansion coefficients $\alpha_{\perp}$ and $\alpha_{\parallel}$ is expressed in terms of the effective Gruneisen functions $\bar{\gamma}_{\perp}^{\ell}(T)$ and $\bar{\gamma}_{\parallel}^{\ell}(T)$ as follows:

$$V\alpha_{\perp} = \left[(S_{11}+S_{12}) \, \bar{\gamma}_{\perp}^{\ell}(T) + S_{13} \, \bar{\gamma}_{\parallel}^{\ell}(T) \right] C_V$$

$$= \bar{\gamma}_{\perp}^{Br}(T) \, C_V \, \chi_{iso}$$

$$V \alpha_{||} = \left[2S_{13} \, \bar{\gamma}_{\perp}^{\ell}(T) + S_{33} \, \bar{\gamma}_{||}^{\ell}(T) \right] C_V$$

$$= \bar{\gamma}_{||}^{-Br}(T) \, C_V \, \chi_{iso} \qquad \qquad .. \ (7)$$

Here S_{ij} are the elastic compliance coefficients, V is the molar volume, C_V is the molar specific heat at temperature T and χ_{iso} is the isothermal compressibility. $\bar{\gamma}_{\perp}^{Br}(T)$ and $\bar{\gamma}_{||}^{Br}(T)$ are the average Gruneisen functions used by Brugger and Fritz (11). The low temperature limits of the Brugger gammas, i.e. $\bar{\gamma}_{\perp}^{Br}(-3)$ and $\bar{\gamma}_{||}^{Br}(-3)$ are obtained, by using the calculated values of $\bar{\gamma}_{\perp}^{\ell}(-3)$ and $\bar{\gamma}_{||}^{\ell}(-3)$ for scandium in Eqn.(7), as 0.34 and 0.26 respectively.

Table 2: Wave Velocities and Generalized Gruneisen Parameters for the Elastic Waves in Scandium Propagating at Different Angles θ to the Hexagonal Axis[a]

θ	V_1	γ_1'	γ_1''	V_2	γ_2'	γ_2''	V_3	γ_3'	γ_3''
5	1.89	0.96	2.90	0.97	0.05	2.29	0.96	0.09	2.23
15	1.87	0.93	2.92	0.99	0.15	2.11	0.96	0.48	1.67
25	1.83	0.89	2.92	1.04	0.34	1.77	0.96	1.04	0.87
35	1.80	0.89	2.78	1.09	0.60	1.31	0.96	1.48	0.25
45	1.77	1.05	2.37	1.11	0.90	0.78	0.95	1.59	0.16
55	1.77	1.38	1.66	1.09	1.20	0.25	0.95	1.29	0.66
65	1.78	1.75	0.92	1.05	1.46	−0.23	0.95	0.75	1.44
75	1.80	2.02	0.38	1.00	1.66	−0.59	0.95	0.18	2.19
85	1.82	2.17	0.11	0.97	1.77	−0.78	0.95	−0.19	2.65

[a] The velocities are in units of $10^{11/2}$ cm/sec.

The low temperature limit of the volume lattice thermal expansion of scandium is given by

$$\bar{\gamma}_L = 2\,\bar{\gamma}_\perp^{Br}(-3) + \bar{\gamma}_\parallel^{Br}(-3) = 0.94 \qquad .. (8)$$

The Temperature Dependence of the Effective Gruneisen Functions

The procedure of Blackman (12) is adopted to calculate $\bar{\gamma}_v^\ell(T)$ as a function of temperature. The normal mode frequencies $\omega(\vec{q},j)$ and the individual GPs $\gamma'(\vec{q},j)$ and $\gamma''(\vec{q},j)$ have been calculated using a computer program (IBM 370/155), at 84 points evenly distributed over 1/24th of the volume of the Brillouin zone. (This is equivalent to 1060 points in the total volume of the Brillouin zone). The range of frequencies from zero to ω_{max} is divided into small intervals ($\Delta\omega = 0.3 \times 10^{13}$ Hz) and the number of frequencies in each interval is counted. A histogram is constructed for $g(\omega)$ vs ω and is replaced by a smooth curve enclosing unit area with the ω-axis. This gives the normalised frequency distribution curve for scandium and is shown in Figure 1.

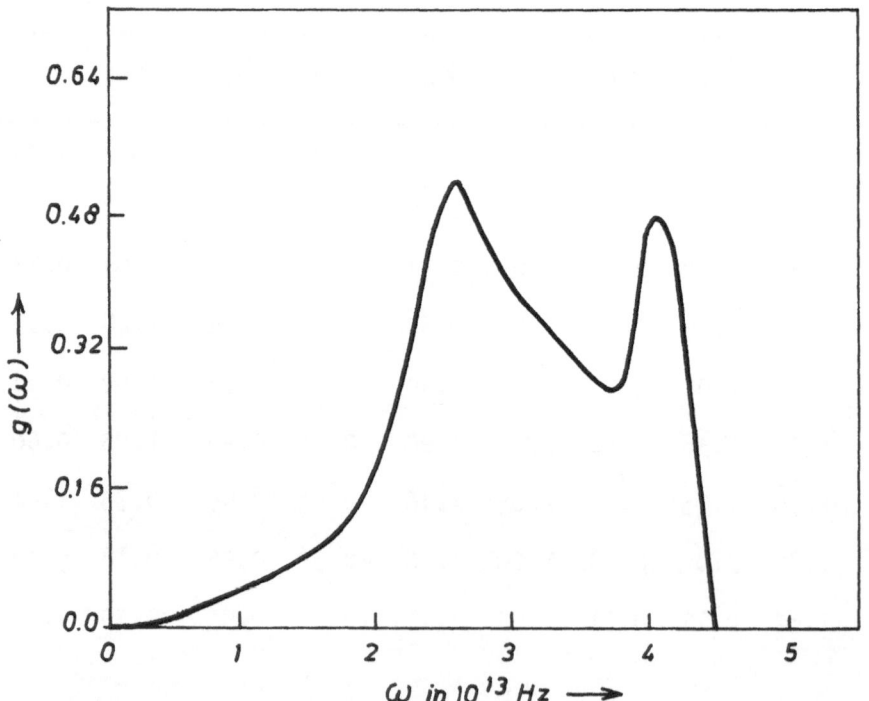

Fig.1: Normalized frequency distribution function for Scandium

In the low frequency region the distribution curve is parabolic, according to the equation $g(\omega) = C\omega^2$. The constant C is determined from the average value of $\sum\limits_{j=1}^{3} V_j^{-3}(\theta,\phi)$ over all directions (θ,ϕ), where $V_j(\theta,\phi)$ is the acoustic wave velocity of the jth mode propagating in the (θ,ϕ) direction. The even moments of the frequency distribution function for scandium on the present model have been determined and are

$\mu_2 = 10.15 \times 10^{26} \text{ sec}^{-2}$, $\mu_4 = 127.91 \times 10^{52} \text{ sec}^{-4}$,

$\mu_6 = 1832.4 \times 10^{78} \text{ sec}^{-6}$. The high temperature limit of the

Debye temperature has been calculated using the value of μ_2 and the formula

$$\theta_\infty = (h/k_B)(5\mu_2/3)^{\frac{1}{2}} \qquad\qquad .. \ (9)$$

Its value for scandium is found to be 314° K.

The individual GPs γ' and γ'' for the various normal modes in each frequency interval ($\Delta\omega = 0.3 \times 10^{13}$ Hz) are noted and the average values of these GPs $\bar{\gamma}'$ and $\bar{\gamma}''$ are calculated for each interval. The $\bar{\gamma}'(\omega)$ and $\bar{\gamma}''(\omega)$ versus ω curves are shown in Fig.2.

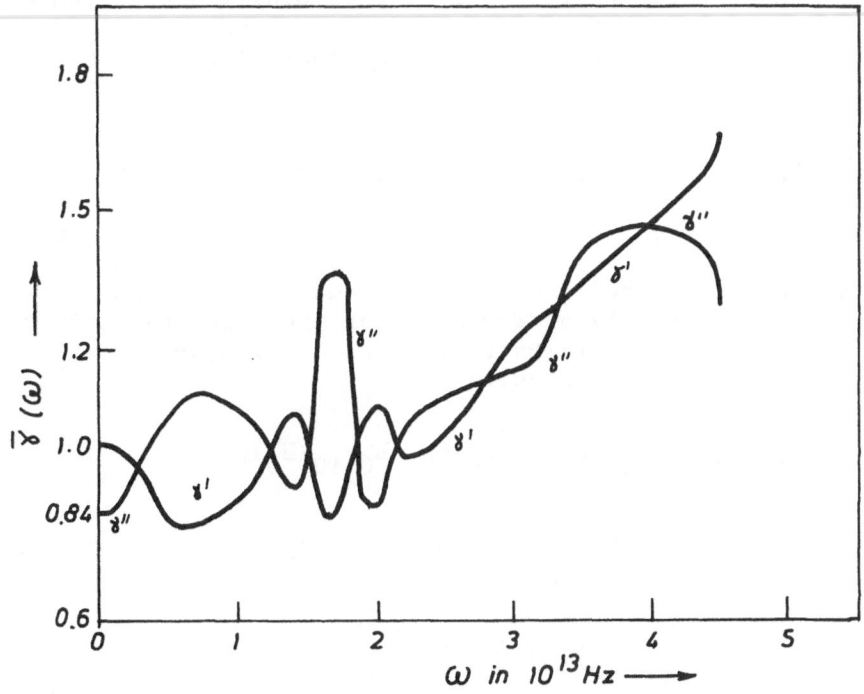

Fig.2: $\bar{\gamma}'(\omega)$ and $\bar{\gamma}''(\omega)$ vs ω for Scandium

In Fig.2 the low frequency range $\bar{\gamma}'(\omega)$ and $\bar{\gamma}''(\omega)$ tend to their respective low temperature limits 1.0 and 0.84. The effective Gruneisen functions are then calculated as follows:

$$\bar{\gamma}_{\perp}^{\ell}(T) = \int_{0}^{\omega_{max}} \bar{\gamma}'(\omega)\, g(\omega)\, \sigma(\omega,T)\, d\omega \,\Big/\, \int_{0}^{\omega_{max}} g(\omega)\, \sigma(\omega,T)\, d\omega$$

$$\bar{\gamma}_{\parallel}^{\ell}(T) = \int_{0}^{\omega_{max}} \bar{\gamma}''(\omega)\, g(\omega)\, \sigma(\omega,T)\, d\omega \,\Big/\, \int_{0}^{\omega_{max}} g(\omega)\, \sigma(\omega,T)\, d\omega$$

$$\cdot\cdot\,(10)$$

Here $\sigma(\omega,T)$ is the Einstein specific heat function. The corresponding Brugger gammas at each temperature are calculated using Eqn.(7) and hence the volume gamma is obtained using

$$\bar{\gamma}_{V}^{\ell}(T) \;=\; 2\,\bar{\gamma}_{\perp}^{Br}(T) \;+\; \bar{\gamma}_{\parallel}^{Br}(T) \qquad\qquad \cdot\cdot\,(11)$$

The temperature dependence of $\bar{\gamma}_{V}^{\ell}(T)$ is shown in Fig.3. The effective volume Gruneisen function $\bar{\gamma}_{V}^{\ell}(T)$ shows a minimum around 7°K and thereafter increases rapidly until 180°K. It then becomes nearly flat and finally attains the high temperature limit of 1.22. This agrees well with the value 1.17 reported by Gschneidner obtained from thermal expansion and C_V^{ℓ} data.

VARIATION OF THE LATTICE PARAMETERS WITH HYDROSTATIC PRESSURE

The theoretical TOE constants of this model have been used in Thurston's (6) formula to calculate the changes in the lattice parameters "c" and "a" of scandium due to hydrostatic pressure. The formula of Thurston for the principal stretches λ_i (i = 1,2,3) which is consistent with a linear pressure dependence of the bulk modulus is

$$\lambda_i = \left[(B/B_0)^{-B_0^2\, y_{i0}/(B_0')^2} \right] e^{\,(a_i + B_0 y_{i0}/B_0')p} \qquad \cdot\cdot\,(12)$$

For a uniaxial crystal $\lambda_1 = \lambda_2 = \lambda_{\perp}$ and $\lambda_3 = \lambda_{\parallel}$ Hence Equation (12) becomes

$$\lambda_{\perp} = \; a/a_0 = \left[(B/B_0)^{-B_0^2 y_{\perp 0}/(B_0')^2} \right] e^{\,(a_{\perp} + B_0 y_{\perp 0}/B_0')p}$$

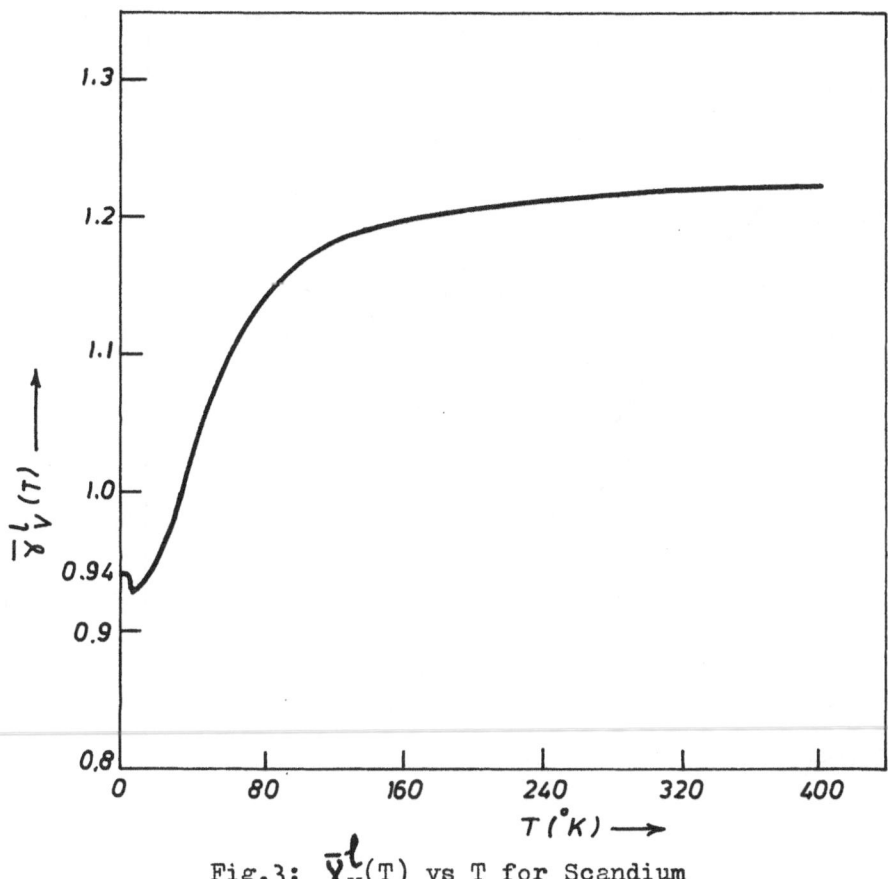

Fig.3: $\bar{\gamma}_V^{\ell}(T)$ vs T for Scandium

and $\lambda_{\parallel} = c/c_0 = \left[(B/B_0)^{-B_0^2 y_{\parallel 0} /(B_0')^2}\right]$ e $^{(a_{\parallel} + B_0 y_{\parallel 0} /B_0')p}$

$$\qquad\qquad .. \ (13)$$

Here a_0 and c_0 are the lattice parameters at zero pressure, while a and c are the same at a pressure p . B is the bulk modulus at a pressure p . B_0 and B_0' are the bulk modulus and the pressure derivative of the bulk modulus at zero pressure. In Eqn.(13)

$$B_0 = -1/(2a_{\perp} + a_{\parallel}), \qquad B_0' = B_0^2 (2y_{\perp 0} + y_{\parallel 0})$$

$$y_{\perp 0} = b_{\perp} - a_{\perp}^2 \quad , \qquad y_{\parallel 0} = b_{\parallel} - a_{\parallel}^2 \qquad .. \ (14)$$

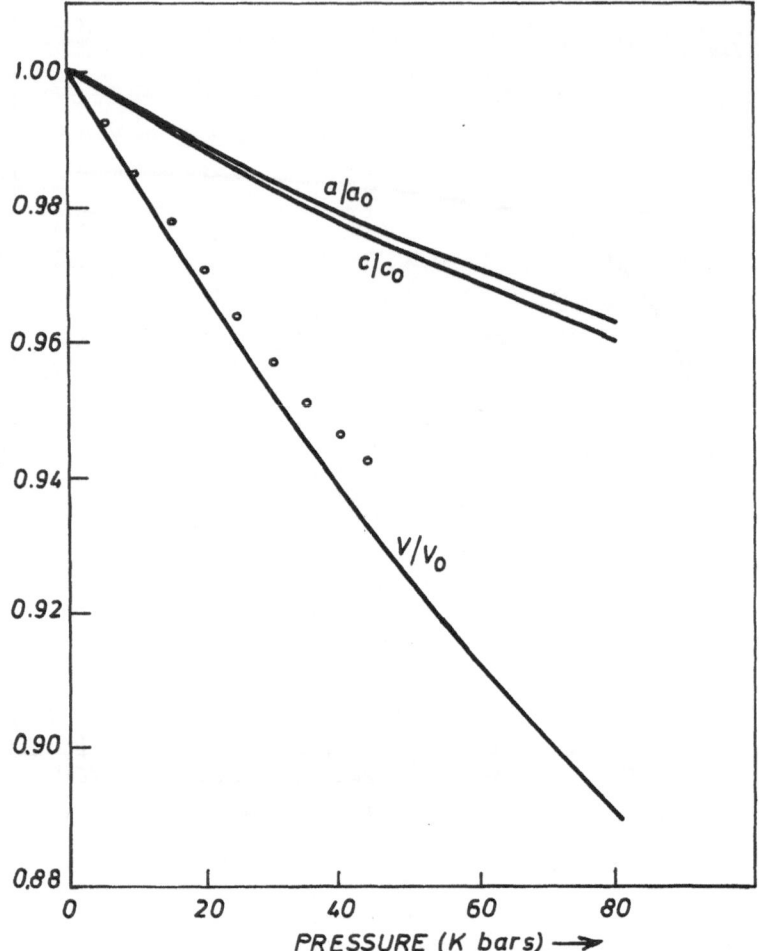

Fig.4: Variation of the lattice parameters and volume
of Scandium with pressure. The dots are experimental points.

where
$$a_\perp = -S_1 = -(S_{11} + S_{12} + S_{13})$$

$$b_\perp = S_1^2 + (S_{11} + S_{12})Q_1 + S_{13}Q_3$$

$$a_{||} = -S_3 = -(2S_{13} + S_{33})$$

$$b_{||} = S_3^2 + 2S_{13}Q_1 + S_{33}Q_3 \qquad\qquad \cdots (15)$$

Table 3: Values of the Parameters used in the present
          calculations of the compression of Scandium

$a_\perp$ $(\text{Mbar})^{-1}$	$a_\parallel$ $(\text{Mbar})^{-1}$	$b_\perp$ $(\text{Mbar})^{-2}$	$b_\parallel$ $(\text{Mbar})^{-2}$	B_0 (Mbar)	B_0'
−0.563	−0.627	3.200	4.222	0.5705	3.123

In Eqn.(15)

$$Q_1 = S_1 (\beta_{11}' + \beta_{12}') + S_3\beta_{13}',$$

$$Q_3 = 2S_1\beta_{13}' + S_3\beta_{33}' \qquad \qquad .. (16)$$

Here S_{ij} are the elastic compliance coefficients and the β_{ij}' are the pressure derivatives of the effective elastic coefficients (Table 1). From Eqn.(13) the volume ratio is

$$V/V_0 = (\lambda_\perp)^2 \lambda_\parallel = (1 + pB_0'/B_0)^{-1/B_0'} \qquad .. (17)$$

Eqn.(17) is the equation derived first by Murnaghan (13).

The values of a/a_0, c/c_0 and V/V_0 for scandium at different pressures upto 80 K.bar have been calculated using Eqns. (13) and (17) and the theoretical pressure derivatives of the elastic coefficients. The values of the parameters $a_\perp$, $a_\parallel$, $b_\perp$ and $b_\parallel$ and those of B_0 and B_0' for scandium are given in Table 3. The variation of the lattice parameters and volume with pressure is shown in Fig.4. The agreement between the calculated V/V_0 values and those measured by Stephens (3) is good (within 1%).

DISCUSSION

The nearly equal magnitudes of the three T.O.E. constants C_{111}, C_{222} and C_{333} indicate that there is higher order elastic isotropy in scandium. The GPs in this metal (both γ' and γ'') have small values characteristic of the hcp rare earth metals. The pressure derivative $\partial C_{44}/\partial p$ of scandium is positive, unlike in zirconium (14), from which it may be inferred that the pressure-induced phase transformation from hcp to bcc structure is not possible in this metal. The calculated $\bar{\gamma}_H$ value agrees well with the value reported by Gschneidner. There seems to be no low temperature thermal expansion data for this metal and hence comparison of $\bar{\gamma}_L$ with experiment is not possible.

The volume compression ratios V/V_0, calculated using the theoretical TOE constants are in close agreement with experiment. This speaks well for the TOE constants of this model. We are thus led to the expectation that the presently calculated TOE constants should show good percentage agreement with any future experimental measurements.

ACKNOWLEDGEMENT

One of the authors (A.R.) is grateful to the Council of Scientific and Industrial Research, Government of India for the award of a research fellowship.

REFERENCES

1.   E.S.Fisher and D.Dever, Proc. Seventh Rare Earth Research Conference, Vol.1, 237 (1968)
2.   K.A.Gschneidner Jr. in Solid State Physics 16, F.Seitz and D.Turnbull, eds.(Academic Press, New York,1964),p.412.
3.   D.R.Stephens, J.Phys.Chem.Solids,25, 423(1964)
4.   R.Srinivasan and R.Ramji Rao, in Inelastic Scattering of Neutrons (IAEA, Vienna,1965), Vol.1, p.325.
5.   R.Ramji Rao and R.Srinivasan, Phys.Stat.Solidi,29,865(1968)
6.   R.N.Thurston, J.Acoust.Soc.Am.41, 1093(1967)
7.   R.Ramji Rao and A.Ramanand, J.Low Temp.Phys.26,365 (1977)
8.   R.Ramji Rao and R.Srinivasan, Phys.Stat.Solidi.31,K39(1969)
9.   F.D.Murnaghan, Finite Deformation of an Elastic Solid (Wiley, New York, 1951)
10. R.Ramji Rao and R.Srinivasan, Proc.Ind.Nat.Acad.Sci.A36, 97(1970)
11. K.Brugger and T.C.Fritz, Phys.Rev.157, 524 (1967)
12. M.Blackman, Proc.Phys.Soc.(London) B70, 827(1957)
13. F.D.Murnaghan, Proc.Nat.Acad.Sci 30, 244(1944)
14. E.S.Fisher, M.H.Manghnani and T.J.Sokolowski, J.Appl. Phys.41, 2991(1970).

THERMAL EXPANSION OF METALS OVER THE ENTIRE LIQUID RANGE

J. W. Shaner

University of California Lawrence Livermore Laboratory

Livermore, California 94550

ABSTRACT

This paper reviews the current state of the art for measuring liquid metal densities. Conventional high precision techniques for use below 2000K as well as new techniques for more extreme temperatures are addressed. Pertinent data, which have appeared since the last critical reviews, for elemental metals are discussed.

I.   INTRODUCTION

Thermal expansion data requires the measurement of two quantities: a length or density, and a temperature. For solids, high precision measurements of both of these quantities can be made, at least up to modest temperatures (T ∿ 1000K). The precision available from optical interferometry or capacitance dilatometers, and thermocouples or resistance thermometers can result in ±0.1% accuracies in thermal expansion coefficient around 300K.[1,2] In fact, the precision may be so high that the effects of even 0.1% Fe impurity on the thermal expansion of ZnS have been detected.[3]

As the temperature is increased, both length or density, and the temperature become more difficult to measure with precision. Very careful measurements of the linear expansion coefficient of Mo and W wires up to 2800K and 3500K respectively have been made with reported accuracies of ±1%.[4] More recently, an optical interferometer has been used with subsecond pulse heating of metals up to their melting points.[5] This technique has promise of also giving ±1% measurements of linear expansion coefficients up to

69

2000–3000K. Modulation techniques[6] can be exploited to give direct measurement of expansion coefficients, thereby eliminating the errors inherent in differentiating density or length vs temperature data. One might hope that these techniques will become accurate enough to measure non-equilibrium effects associated with vacancy concentration near the melting point.

The theoretical situation for thermal expansion of solids is also relatively well developed. This is particularly true for ionic crystals like alkali halides or molecular crystals where interparticle potentials are known well enough to make realistic calculations of anharmonic effects.[7] For simple metals, at near normal density, the pseudopotential approach can be used to define effective interparticle potentials, which can then be used to evaluate the thermodynamic functions.[8]

Thermal expansion measurements on liquid metals present several problems not normally encountered in measurements on solids. Firstly, the temperatures are almost always high. Secondly, the samples are not self-supporting. Therefore either a container is necessary, or the heating must be done rapidly while inertially confining the liquid. Thirdly, if a container other than an a.c. magnetic field is used, the extreme corrosiveness of most liquid metals must be considered.

On the other hand, the problem of non-equilibrium defect concentrations may not apply to liquid metals. Therefore, dynamic measurements, which may avoid the container problems, are not so suspect as they are in the solid phase near melting.

Theoretical calculations of the thermal expansion of liquids can be accomplished through differentiation of the equation of state. A standard result of statistical mechanics is that, assuming pairwise additive spherically symmetric potentials, the equation of state can be written as[9]

$$P = \frac{N}{V} kT - \frac{2\pi}{3} \frac{N^2}{V} \int_0^\infty \frac{d\phi(r)}{dr} g(r) r^3 dr \qquad . \qquad (1)$$

In this equation, $\phi(r)$ is the interparticle potential and $g(r)$ is the radial distribution function. For simple liquids, like argon, the interparticle potential is known well enough to calculate a reliable equation of state with even a hard sphere distribution function. However, for metals the pseudopotential, $\phi(r)$, is density dependent, so calculations become much less reliable. Pseudopotential calculations of the thermal expansion coefficient of even simple liquid metals at the melting point have come only within 30–40% of experimental values.[10]

In view of the uncertainties inherent in a direct calculation of liquid metal equations of state, accurate measurement is desirable for many applications. For example, the volume change upon melting is an important parameter for evaluating residual stresses in weld joints. Also, the equation of state is necessary for designing for accidents in nuclear reactors. For this latter application, equation of state for some materials must be known up to ∿1 GPa in pressure and ∿10,000K in order to properly model some advanced reactors.[11]

The thermal expansion of liquid metals has been the subject for several very extensive reviews covering the period up to 1973.[12-17] Since the conventional techniques for measuring liquid metal densities up to 2000K are described in detail in these reviews, these techniques will be only briefly discussed in this paper. New data appearing after 1973 will be described along with new techniques capable of significantly increasing accuracy or temperature range.

II. TECHNIQUES REQUIRING A CONTAINER

Perhaps the most familiar apparatus for measuring thermal expansion of a liquid metal is the mercury thermometer. This device is the prototype of all dilatometers, where the liquid metal density is measured directly as a change in surface level with temperature. At temperatures below 1700K, the containing vessel may be quartz for many metals, and the liquid surface may be viewed directly. Under these circumstances the surface level may be corrected for the meniscus shape. For work at higher temperatures a refractory container must be used--tantalum or graphite, for example. Then typically the liquid level is monitored by a contacting electrode. The difficulties in this technique center around calibrating the relation between surface level and sample volume and accurately measuring surface level in spite of optical aberrations, surface curvature, and dilatometer expansion. Since the whole apparatus can be set in a furnace, the temperatures can be accurate to better than 1 K. However, to make density measurements better than ±0.5% at temperatures greater than 1300K is practically impossible.[16]

A potentially very accurate capacitance technique for monitoring the liquid level in a zirconia tube has been used by Evan and Magen to obtain density of liquid mercury up to 1800K.[18] The tube has been metallized on the outside to form one electrode of the capacitor, while mercury metal inside forms the other electrode. For mercury, this tube was maintained at room temperature, so no corrections for temperature dependent surface tension (i.e. meniscus shape) need to be made. The dilatometer was connected to the heated mercury sample volume by a capillary of negligible volume. For

other metals, the dilatometer would have to be maintained above
the sample melting temperature. The capability for remote sensing
with no movable parts permit this apparatus to be used in a pressure vessel (up to 0.2 GPa in the work of Evan and Magen). Evan
and Jortner, using this apparatus quote an absolute density uncertainty of 1%.[19]

Another of the classical techniques for liquid metal density
measurement is the Archimedean method. In this experiment a sinker
of known mass made of some refractory material which will not be
attacked by the liquid sample is weighed as it is immersed in the
liquid metal. An indirect technique, where the sample is in the
sinker, which is immersed in an inert liquid of known density has
also been applied. In both of these techniques one major source
of error is the force acting on the supporting wire as it passes
through the melt surface. Both the liquid surface tension and the
contact angle must be calibrated. In addition, the thermal expansion of the sinker must be known accurately. Grosse et al. have
used the direct Archimedean method on liquid uranium up to 1900K
with a quoted accuracy of ±0.08%.[20] However, subsequent measurements have given a liquid density smaller by 3%.[21] More recent
measurements on several metals by Lucas[13,14] have been more carefully calibrated for the surface tension effect. These measurements quote uncertainties up to ±0.5% at 2000K, which are probably
realistic estimates of the accuracy of this technique.

Pycnometry, another classic method, involves the filling of a
known volume with liquid metal at a known temperature, and subsequently measuring the mass of the metal after it has cooled. This
technique, thoroughly reviewed by Crawley[16], has a potential
accuracy of better than ±0.05% in density below 750K. This precision cannot, however, be obtained without extreme care to avoid
bubbles or voids when liquid metal fills the cavity.

The last of the classic methods for measuring liquid metal
density is the maximum bubble pressure technique. In this method,
the pressure necessary to force inert gas through a capillary to
form bubbles beneath the surface of the melt is measured. The
change in pressure as the depth of the bubbling capillary tip is
varied measures the hydrostatic head. Corrections which must be
made include the thermal expansion of the capillary and the change
in melt level as the capillary is immersed. A detailed analysis
of the uncertainties in this technique has been given by Lucas[12],
with the result that, at least at modest temperatures, the density
can be measured to ±0.2%. For higher melting temperature metals
an accuracy of ±0.5 - 1% is more appropriate.[22]

For all of the methods described in this section attention
must be given to the possibility of corrosion of the container and

contamination of the sample. If a correction involving surface
tension must be made, then contamination and oxidation of the surface become particularly important. Temperatures can be measured
accurately by thermocouples up to 2000K, but care must be taken
to prevent contamination of the junction. Optical pyrometry can
be used if ±5-10 K uncertainties can be tolerated, but again care
must be exercised if the metal sample has a high vapor pressure.
With all of these cautions in mind, it seems possible to obtain
densities accurate to ±0.5% at temperatures up to 2000K by these
classic techniques.

III. DROP TECHNIQUES

A drop of a liquid metal can be made to bead on a hot sub-
trate (typically of graphite), it can be allowed to hang or drop
from a rod of the same material, or from a tube, and a drop can
be levitated by an a.c. magnetic field. Conventionally the volumes
of the drops are measured photographically in either visible light
or with x-rays. For static drops the shapes are determined by
both surface tension and density, so both parameters can be
measured. On the other hand, the accuracy of the density is less
because the drop is not spherical. If the drop is photographed
during free fall, the shape should be spherical unless a large
amplitude hydrodynamic oscillation has been excited. However,
the photography is more complicated. The uncertainties inherent
in drop techniques have been summarized by Saito et al.[22] They
found the total uncertainty to be ±2% in density up to 2000K, with
the measurement of the drop image the largest error. The temperatures in these experiments must be obtained pyrometrically, and
this contributes extra uncertainty to the measurements.

Recently double pulse holography has been applied to obtain
directly the thermal expansion coefficient of a levitated aluminum
drop.[23] The holography allows dimensional changes of less than
1 μm to be determined, but the drop size must still be measured
photographically. The volume expansion coefficient obtained in
this work, 1.95×10^{-4} K^{-1} at 1100K agrees reasonably with a value
of 1.65×10^{-4} chosen by Crawley as the most reliable measurement.[16]
The temperature of the levitated sample is difficult to control,
and it must be changed rapidly in order to not lose material by
evaporation. Also the position of the drop must be kept stable to
an optical wavelength. As a result, the double pulse holography
may remain just a curiosity in the field of liquid metal density
measurements.

Although the ±2% density measurements made with drop techniques
are not the most accurate, drop methods have allowed measurements at
higher temperatures. In particular, Russian researchers have

exploited drop techniques to obtain liquid densities at the melting point of refractory 4d and 5d transition metals.[24-28] The reason for the high temperature capability is that a container is unnecessary.

IV.   RADIATION ATTENUATION METHOD

The attenuation of gamma rays upon passing through a liquid metal specimen of known thickness has been applied with success to the measurement of density. The highest accuracies are obtained by counting many transmitted gamma rays, but the incident beam cannot be so intense that it disturbs the temperature distribution. As a result, except for very crude measurements this technique will continue to be used with static samples in a container. This restriction will limit the upper temperature range to 2000-3000K. The technique is described in more detail by W. Drotning at this conference.

V.   DIRECT TECHNIQUE

A recently reported experimental technique, the isobaric expansion experiment (IEX)[29], has been used by us to obtain liquid metal densities over a very extensive temperature range.[30-32]

In essence, our experimental procedure consists of resistively heating a 1 mm diameter 25 mm long metal wire sample in an inert gas filled pressure vessel. The inert gas provides an isobaric environment, since the gas volume is much larger than the sample volume, and the sample expansion is slow enough to allow acoustic reverberations in the cell. The inert gas also prevents chemical contamination of the hot wire surface, an important feature for our optical pyrometry. In addition, the high pressure raises the boiling temperature, allowing us to investigate most of the liquid metal density range.

Heating of the sample is accomplished by discharging a critically damped capacitor bank through the sample quickly enough to overcome the radiative heat losses and hydrodynamic instabilities (t < 100 μs), but slowly enough so that the skin effect doesn't cause serious energy inhomogeneities (t > 10 μs). Typically, a roughly square current pulse of 30 kA with 5 μs rise and fall times is used. With these current pulse parameters the energy density is homogeneous to within 1% by the time the sample is heated to 2000K, where the pyrometry starts responding.

The pressure vessel is designed with four sapphire windows viewing the sample in a radial direction. Back lit shadowgraphs

are taken both with a Q-switched ruby laser and with a cw laser
and streaking camera. Since the metal is constrained at the ends
by clamping jaws, its expansion in the liquid phase is only radial.
Therefore the square of the diameter, measured as a function of
time by the streaking camera, is proportional to the sample specific
volume. The volume measurement can be made to ±2%. The ruby laser
snapshot is taken typically 5-10 μs after the end of the current
pulse. If the final state of the sample is not uniform along the
axis, the shot is discounted.

The current is measured with a wide-band current transformer,
and the voltage drop across the middle 7-10 mm of the wire is
measured in a separate high impedance circuit. The oscilloscope
traces of both of these signals are digitized, and, after a slight
inductive correction is made, the product of current and voltage
is integrated to give an enthalpy vs time curve. The enthalpy so
obtained is accurate to ±3%. This is correlated with volume vs
time to give an equation of state.

In order to make pyrometric records, a patch of the sample,
approximately 1 x 3 mm, is viewed through a randomly trifurcated
fiber optic bundle. The three output bundles deliver the thermal
radiation through 50 mm band pass filters (typically at 450, 700,
and 900 nm) to silicon photodiodes. The edges of the sample are
carefully avoided in the image viewed by the pyrometer, since deviations from Lambert's law have been observed for thermal radiation
coming off at grazing angles. The photodiode signals are amplified
logarithmically and displayed on oscilloscopes. The log amplifiers
reduce the precision of our light measurements, but they allow a
whole shot covering the 2000 to 8000K temperature range to be displayed on a single trace. The generally unknown wavelength and
temperature dependences of liquid metal emissivities make the calculation of temperatures from the pyrometer readings something of
an art. The details of the calculation of temperatures from the
pyrometric records is given in reference 31. The temperatures are
estimated to be accurate to ±200K at 5000K.

The IEX technique, like all the others, has both good and bad
features. The density measurements are only accurate to ±2% because of limitations of the measurement technique and difficulties
in maintaining a smooth surface at extreme temperatures in a dynamic
experiment. Also, the temperature measurement is inaccurate because radiation pyrometry must be used in a range where emissivities
are generally unknown. On the other hand, densities of metals can
be measured at much higher temperatures than was previously possible.
For some simulation applications, the enthalpy is a more convenient
variable than temperature, and enthalpy is measured reliably in IEX.
Finally, even the most corrosive liquid metals may be studied.

VI. NEW DATA

In the four years since the last critical reviews[16],[17], there
have appeared several new determinations of liquid metal densities
in temperature ranges already explored, as well as new measurements
at higher temperatures.

Bismuth and Zinc have been remeasured in the temperature
ranges 544-900 and 693-900K respectively by the Archimedean
method.[33] In both cases, the data can be represented as
$\rho = a + b(T-Tm)$, and the liquid densities at the melting point, a,
differ by less than 0.3% from the values preferred by Crawley.[16]
The values of b differ from Crawley's values by 3% for Zn and 1%
for Bi. Dilatometric measurements of the densities of liquid Cd,
In, Sn, and Sb, and their binary alloys have been made by Nakajima[34]
with agreement with Crawley's preferred values of a to ±0.2% in
all cases. Larger discrepancies are found in Nakajima's b param-
eters--as large as 10%. This may be due to the very limited temperature range over which his data has been taken (200-300K).

Gold[35], tin[35], and silver[36] have been measured by sessile
drop techniques with results agreeing with Crawley to better than
1% for a. The values of b differ from Crawley's by as much as 70%.

The differences between these new measurements and previous
determinations probably represent the accuracy one can expect in
the density measurements--i.e. ±0.1-0.3% below 1000K and 1% below
2000K. If density change is measured over a 1000K temperature
range, ±0.1% uncertainties in a result in ±10% in b. For smaller
temperature intervals, the uncertainties in b are even greater.

Copper and iron have been subject to many density measurements
in the last few years. These results are summarized in Tables I
and II. Discrepancies of 1% in density and 10-20% in b still
appear for these common metals at temperatures above 1500K.

Recent isobaric expansion data for liquid lead are presented
in Figure 1. To within experimental error, a straight line fit
to the data of the form $\rho = 10.6 - 11.6 \times 10^{-4} (T-600)$ gm/cm^3 fits
the data over a twofold density change. Agreement with other measurements below 2000K is satisfactory. The thermal expansion coefficient obtained here is slightly smaller than those recommended
by Crawley[16] and Steinberg.[17] This may be due to the 0.2 GPa
ambient pressure used in the IEX experiments.

Similar data is shown for uranium at 0.2 GPa in Figure 2. In
this case the fit to the IEX data is $\rho = 17.6 - 17 \times 10^{-4} (T-1406)$
gm/cc. Since the temperature range of these experiments is very
large, the uncertainty in the b coefficient is the same magnitude
as the uncertainty in the a coefficient--±2%.

TABLE I

LIQUID COPPER DENSITY

a(gm/cc)	−bx10⁴	Temp. Range(K)	Technique	Ref.
7.936	7.862	1356–1512	Archimedean	37
8.019	7.3	1356–1873	γ-Attenuation	38
7.940	7.367	1376–1556	Pycnometry	39

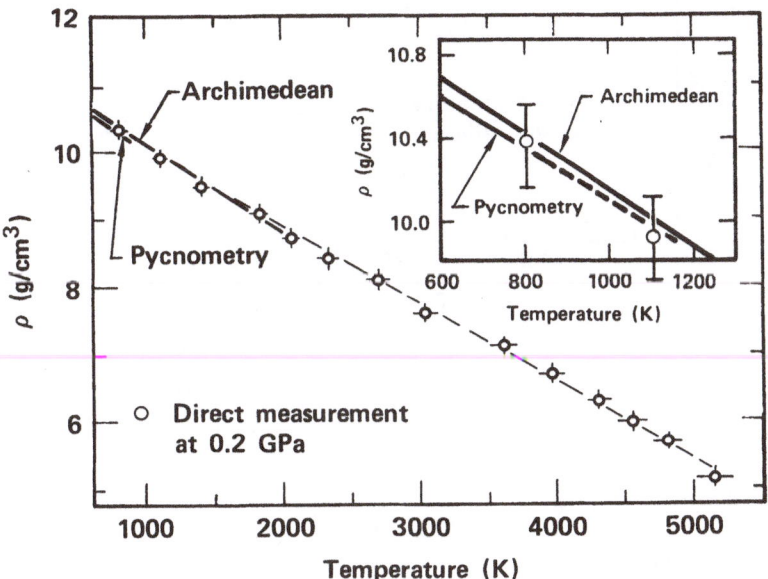

Figure 1 Thermal expansion of liquid lead. Archimedean technique
         ref. 42; Pycnometry, ref. 13; Direct measurement, ref. 31.

TABLE II

LIQUID IRON DENSITY

a(gm/cc)	−bx10⁴	Temp. Range(K)	Technique	Ref.
7.078	6.5	1812–1993	γ-Attenuation	38
7.2	10.4	1833–2023	Dilatometry	40
7.05	6.1	1810–1923	γ-Attenuation	41

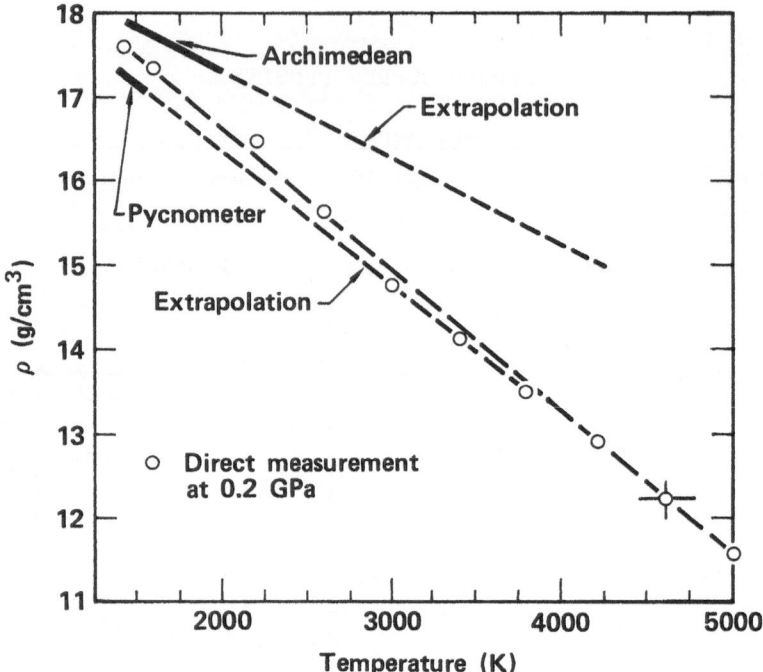

Figure 2   Thermal expansion of liquid uranium.   Archimedean
technique, ref. 20; Pycnometry, ref. 21.

TABLE III

Densities of Refractory 4d and 5d Metals

Metal	Melting Pt. (K)	Density at M.P.	bx10⁴	Temperature Range (K)
Zr	2123	6.09 (24)		
Nb	2741	7.68 (30)	5.4 (31)	(2741-4200)
		7.57 (27)		
Mo	2880	9.04 (30)	8.1 (31)	(2880-5000)
Rh	2239	12.20 (26)	50 (26)	(2239-2473)
Hf	2216	11.97 (24)		
Ta	3270	14.43 (30)	13 (31)	(3270-7000)
W	3690	16.26 (30)		
		16.67 (28)		
Ir	2716	19.39 (25)		

A number of determinations of high melting point 4d and 5d liquid metal densities have recently appeared. These have been obtained by both drop techniques at or near the melting point[24-28], and IEX over an extensive temperature range.[30-31] These data are summarized in Table III.

VII. SUMMARY

Density data for liquid metals has been obtained at precisions of better than ±0.2% below 1000K, ±1% below 2000K and ±2% over the entire liquid range. The IEX technique has provided a useful tool for extending the data base to extreme temperatures. Although the density measurements by this latter technique are less accurate than those obtained by conventional methods, the very large temperature range permits more accurate thermal expansion coefficients to be measured.

References

1. F. R. Krueger and C. A. Swenson, J. App. Phys. 48, 853 (1977).

2. K. G. Lyon, G. L. Salinger, C. A. Swenson, and G. K. White, J. App. Phys. 48, 865 (1977).

3. F. W. Sheard, T. F. Smith, G. K. White, and J. A. Birch, J. Phys. C: Solid State 10, 645 (1977).

4. V. A. Petukhov and V. Ya Chekhovskoi, High Temp. High Press. 4, 671 (1972).

5. A. P. Muller and A. Cezairliyan, proceedings of the Seventh Symposium in Thermophysical Properties, National Bureau of Standards, Gaithersburg, Md., May 10-12, 1977.

6. Ya. A. Kraftmakher, High Temp. High Press. 5, 645 (1973).

7. See, for example, S. Bijunki and R. J. Hardy, J. Phys. Chem. Sol. 38, 147 (1977).

8. D. C. Wallace, Thermodynamics of Crystals, J. Wiley & Sons, N.Y., 1972, p. 404-427.

9. T. M. Reed and K. E. Gubbins, Applied Statistical Mechanics, McGraw-Hill, N.Y. 1973, p. 179.

10. H. D. Jones, Phys. Rev. A8, 3215 (1973).

11.   L. A. Booth, Nucl. Eng. Design 24, 263 (1973).

12.   L. D. Lucas in Techniques of Metals Research, R. F. Bunshah, ed. Vol. IV, part 2, 1970, p. 219.

13.   L. D. Lucas, Mem. Sci. Rev. Met. 69, 395 (1972).

14.   L. D. Lucas, Mem. Sci. Rev. Met. 69, 479 (1972).

15.   S. D. Veazey and W. C. Roe, J. Mater. Sci. 7, 445 (1972).

16.   A. F. Crawley, Int. Metall. Rev. 19, 32 (1974).

17.   D. J. Steinberg, Metall. Trans. 5, 1341 (1974).

18.   U. Evan and J. Magan, J. Phys. E. Sci. Inst. 7, 902 (1974).

19.   U. Evan and J. Jortner, Phil. Mag. 30, 325 (1974).

20.   A. V. Grosse, J. A. Cahill, and A. D. Kirshenbaum, J. Amer. Chem. Soc. 83, 4665 (1961).

21.   W. G. Rohr and L. J. Wittenberg, J. Phys. Chem. 74, 1151 (1970).

22.   T. Saito, Y. Shiraishi, and Y. Sakuma, Trans. Iron Steel Inst. Japan 9, 118 (1969).

23.   H. Spetzler, M. Meyer, and G. Burcher, High Temp. High Press. 6, 529 (1974).

24.   Yu. N. Ivashchenko and P. S. Martsenyuk, Indust. Lab. 39, 64 (1973). [transl. of Zavod. Lab. 39, 42 (1973)].

25.   P. S. Martsenyuk and Yu. N. Ivashchenko, Ukrain. Khim. Zhur. 40, 431 (1974).

26.   E. L. Dubinin, V. M. Vlasov, A. I. Timofeev, S. O. Safonov, and A. I. Chegodaev, Izv. V. U. Z. Tsvetn. Metall. No. 4, 160 (1975).

27.   Yu. N. Ivashchenko and P. S. Martsenyuk, High Temp. 11, 1146 (1974). [transl. of Teplofiz. Vys. Temp. 11, 1285 (1974)].

28.   P. S. Martsenyuk, Yu. N. Ivashchenko and V. N. Eremenko, High Temp. 12, 1161 (1975) [transl. of Teplofiz. Vys. Temp. 12, 1310 (1975)].

29. G. R. Gathers, J. W. Shaner, and R. L. Brier, Rev. Sci. Inst.
     47, 471 (1976).

30. J. W. Shaner, G. R. Gathers, and C. Minichino, High Temp.
     High Press. 8, 425 (1976).

31. J. W. Shaner, G. R. Gathers and W. M. Hodgson, Proc. of 7th
     Symposium on Thermophysical Properties, May 10, 1977,
     Gaithersburg, Md.

32. G. R. Gathers, J. W. Shaner, and D. A. Young, Phys. Rev.
     Lett. 33, 70 (1974).

33. L. Martin-Garin, P. Bedon, and P. Desre, J. Chim. Phys. 70,
     112 (1973).

34. H. Nakajima, Trans. J. I. M. 15, 301 (1974).

35. G. P. Khilya, Yu. N. Ivashchenko, and V. N. Eremenko, Russian
     Metall. (Metally) #6, 72 (1975) [transl. of Izv. Ak. Nauk
     Metall. #6, 87 (1975)].

36. A. P. Andreyev, V. A. Izmaylov, I. S. Ivakhnenko, and
     V. I. Kashin, Russian Metallurgy (metally) #2, 46 (1974)
     [transl. of Izv. Ak. Nauk Metall. #2, (1974)].

37. M. Gomez, L. Martin-Garin, H. Ebert, P. Bedon, and P. Desre,
     Z. Metallkde. 67, 131 (1976).

38. V. I. Yavoyskiy, A. A. Ezhov, V. F. Kravchenko, V. S. Uskov,
     Yu. I. Nebusov, Yu. A. Chernov, and G. A. Dorofeyev, Russian
     Metallurgy (Metally) #4, 44 (1974).

39. C. O. Ruud, M. T. Hepworth, and J. M. Fernandez, Metall.
     Trans. 6B, 351 (1975).

40. S. Watanabe and T. Saito, Trans. J. I. M. 14, 120 (1973).

41. Yu. N. Karmalin, G. F. Stasyuk, M. I. Gladkov, V. M. Bagin,
     L. S. Etelis, and A. B. Borovskii, Indust. Lab. 39, 739 (1972)
     [transl. of Zavod. Lab. 39, 553 (1972)].

42. A. D. Kirshenbaum, J. A. Cahill, and A. V. Grosse, J. Inorg.
     Nucl. Chem. 22, 33 (1961).

*Work was performed under the auspices of the U.S. Energy Research
and Development Administration under contract no. W-7405-Eng-48.

THERMAL EXPANSION OF MOLTEN MATERIALS BY THE GAMMA ATTENUATION

TECHNIQUE*

William D. Drotning

Sandia Laboratories

Albuquerque, New Mexico 87115

ABSTRACT

An apparatus was constructed for measurement of the density
and volumetric thermal expansion of high temperature molten materials using the gamma attenuation technique. It was found that
significant improvements in overall experimental precision could be
realized by employing an automatic gain control in the linear section of the gamma counting system. This device reduced errors due
to system gain shifts and thereby increased the precision of the
gamma counting. Corrections were also made for the dead time of
the electronics system. A preliminary experimental investigation
of the precision of the method was made by measuring small dimensional steps of a tool steel block at a single temperature. A
fractional length increment $\Delta l/l$ of 1% was determined to 1%, equivalent to a length change precision of 10^{-4} cm/cm. In addition,
the precision of the experimental technique was analytically
examined for both absolute and relative density determinations.
Three analytical expressions used to reduce data for liquid density
determinations were evaluated for their precision. Each allows use
of a different set of input data parameters, which can be chosen
based on experimental considerations. Using experimentally reasonable values for the precision of the parameters yields a similar
resultant density precision from the three methods, on the order
of 0.2%.

* This work supported by the U.S. Department of Energy.

INTRODUCTION

The density and volumetric thermal expansion of high temperature molten materials are necessary for technological needs as
well as of basic scientific interest. An example is the need for
data on molten structural and fuel materials for use in nuclear
reactor safety studies. The gamma radiation attenuation technique
for density determinations of molten materials offers several advantages over other methods at high temperatures. First, the gamma
beam offers a probe which is not in thermal or physical contact
with the liquid. In addition, since a free liquid surface is not
involved in the measurement, a number of problems of other methods
are eliminated, including sample volatilization, formation of
oxide surface films, and corrections for surface tension and liquid/
solid interactions.

Though the technique itself is not new, it has not achieved
widespread use as a high temperature technique. Dillon et al. [1]
and LaVert et al. [2] have reported its use for density measurements of liquid alkali metals, and Döge [3] has used the technique
for density measurements in the Pb-Sn system to 1200 K. The densities of liquid lead, cesium, and gallium were examined with the
gamma method to 1400 K by Basin and Solov'ev [4]. More recently,
the technique has been extended to studies of iron and copper to
1900 K by Yavoyskiy et al. [5]. The technique has also been used
extensively for determining the bulk density and water content of
soils [6,7], and for other technological uses at ambient temperature [8].

In this paper, an experimental apparatus using the gamma attenuation method to 1200 K is described, including the details of the
electronic counting system. Sample size and gamma source choice
considerations are also presented. Next, the basic equation of the
method is presented in forms useful for data analysis, including
both relative and absolute density determinations. Finally, an
analysis of the precision of the method is given.

EXPERIMENTAL APPARATUS

With this technique, the density of a material is determined
from the attenuation which is produced by the passage of a collimated beam of mono-energetic gamma radiation through the material,
whose length and absorption coefficient are independently determined.
A schematic of the experimental arrangement is shown in Fig. 1. A
^{137}Cs 0.662 MeV gamma source (activity 4 Ci) is housed in a lead
vault, which provides both radiological shielding and the initial
collimation for the gamma beam. Collimators of stainless steel,
inside diameter 6.4 mm, direct the beam into the high temperature
tubular furnace. The molten material is contained in a precision-
fabricated crucible, with faces machined parallel to each other
and perpendicular to the beam direction. The crucible is supported

by a ceramic pedestal, both of which are contained in an evacuated
ceramic chamber. The alignment of the crucible and pedestal is
achieved using a He-Ne laser beam passing through the collimators,
prior to lowering of the furnace over the chamber. Emerging from
the furnace, the beam is further collimated (aperture 6.4 mm) before
striking the NaI(Tℓ) scintillation detector, which is shielded in
lead. Steel filters of various thicknesses are inserted into the
beam to establish an appropriate gamma flux for the detection
system.

The gamma detection system is outlined in Fig. 2; except for
the scintillation detector (Bicron 3M3P), the apparatus was manufactured by Canberra Industries, Inc., Meridan, Conn. The detector
consists of a NaI(Tℓ) crystal (76.2 mm diam. x 76.2 mm length),
hermetically attached to a photomultiplier tube window. To achieve
count rate stability over extended time periods, the ambient
temperature of the detector is thermostatically-controlled. The
output pulses are amplified and shaped by the preamplifier (Model
802-9) and amplifier (Model 2010). The amplified pulses pass
through a digital stabilizer (Model 1720), which offers automatic
gain control of the linear electronics. This device was found
necessary to provide system counting stability over long time
periods. The stabilizer compensates for gain shifts in the system

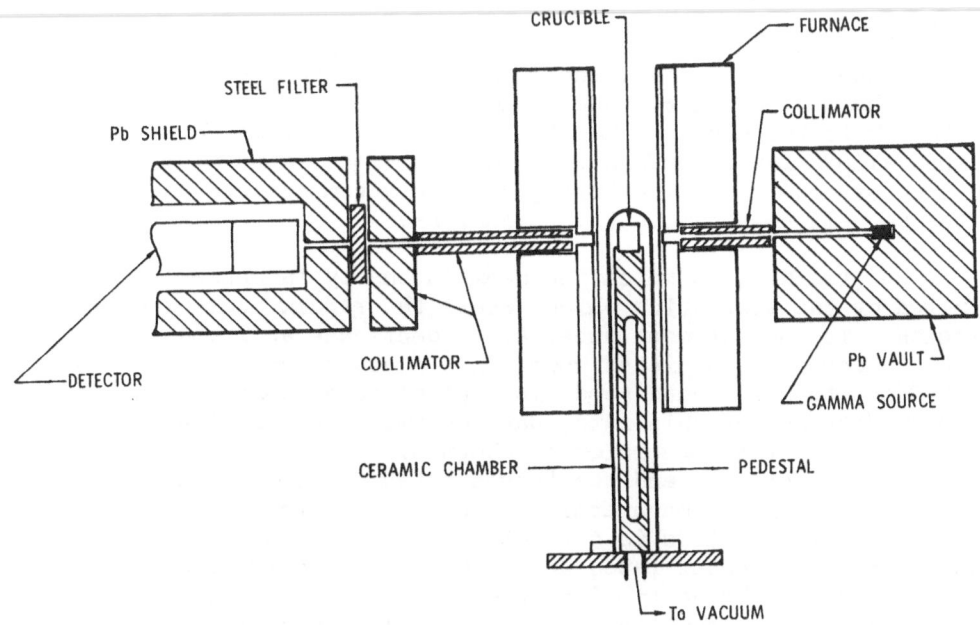

Figure 1. Schematic diagram of gamma attenuation experimental
           apparatus.

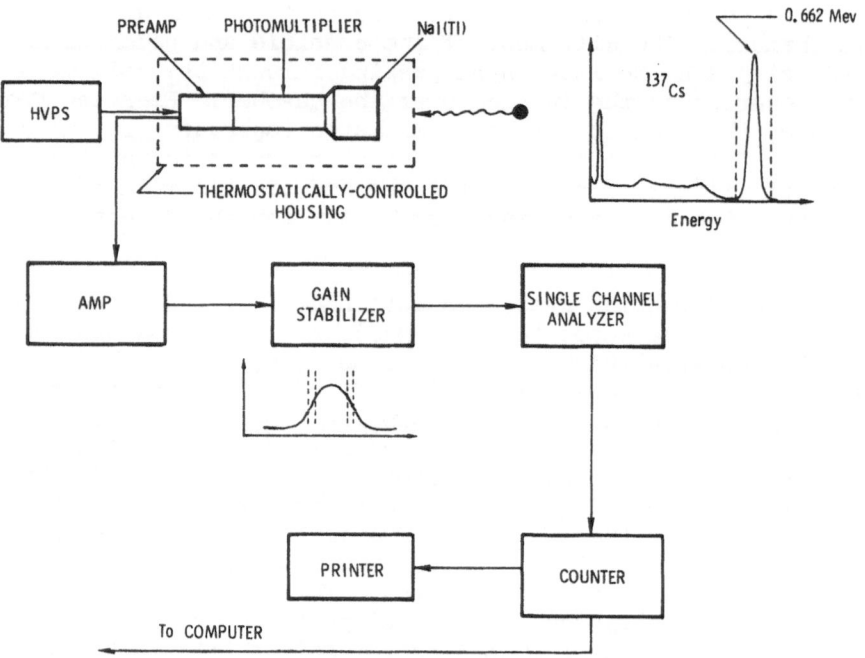

Figure 2. Block diagram of gamma detection electronics.

due to line voltage variations, high voltage power supply variations, count rate changes, and ambient temperature variations. The
single channel analyzer (Model 2030) passes the pulses in the [137]Cs
photopeak to the counter (Model 1790C). The digital format of the
output allows for interface to a computer-based data acquisition
system for continuous monitoring of sample temperature and count
rate, as well as subsequent data analysis.

The precision to which the beam intensity is measured is
limited by the statistics which govern the random nuclear emission
process. The measurement uncertainty decreases as $1/\sqrt{N}$ for a
total number of accumulated counts N. A count rate limit of about
40,000 counts per second (cps) is imposed to accommodate the detection apparatus. At this rate, data accumulation for ten minutes
yields a count rate precision of about 0.02%. Using both the automatic gain stabilizer and the thermostated detector, a stabilized
detection system has been achieved whose count rate precision over
indefinite time periods is determined solely by nuclear counting
statistics. For count rates above 1000 cps, a correction is
required for the "dead time" of the electronics system, which origi-
nates from the probability of a second pulse occurring while the
detection system analyzes a first pulse [9]. This dead time can be
measured for each detection system by the two-source method [10],
or a variation, the two-absorber method [6].

The current apparatus utilizes a resistance-type tubular furnace and graphite crucibles, as shown in Fig. 1, to achieve temperatures to 1200 K. The furnace resistance wires were also drilled out and subsequently bypassed in the beam region. Work is in progress to replace the tubular furnace with a high temperature/high vacuum tungsten mesh furnace capable of 3200 K. In this facility, the gamma beam will be directed through ceramic or refractory metal crucibles and through the electrode gaps in the tungsten mesh heating element.

The basic equation which describes the gamma beam intensity and thus defines the analytical method is [9]

$$I = I_o \exp\left[-u \rho \ell\right] \quad , \tag{1}$$

for mono-energetic gammas and a narrow beam (well-collimated) geometry. In Eq. (1), I_o and I are the gamma intensities before and after passage through the sample material, ρ is the material density, ℓ the material length along the beam direction, and u the mass attenuation coefficient of the material. As the product $(u \rho \ell)$ increases, the attenuation increases, and the uncertainty in ρ due to imprecision in I decreases. On the other hand, as $(u \rho \ell)$ increases, the count rate diminishes, so that longer times are required to yield a given precision; thus, there exists an optimum $(u \rho \ell)$. Considering only uncertainties due to measurement of I and I_o, one can show that, for a fixed total counting time, the fractional standard deviation for ρ is a minimum at $(u \rho \ell) = 2$. (This value also minimizes the counting time necessary to achieve a given precision [7].)

The criterion $(u \rho \ell) = 2$ can be used for selection of a sample length ℓ and the gamma source, since u varies with gamma energy as well as with atomic number Z. To minimize thermal gradients, sample size should not exceed ~ 4 cm in the current furnace. For $\ell = 3$ cm and densities which range from 5-15 g/cm^3, the above criterion gives u $\sim$ .13-.04 cm^2/g. A source is desired which has a long isotope half-life; a relatively simple energy spectrum, giving a strong, monochromatic beam with no higher energy peaks contributing Compton scattered photons; and a high specific activity. As a practical matter, source cost, availability, and shielding requirements eliminate many choices. With these considerations, ^{137}Cs emerges as the best source for most applications. At the ^{137}Cs energy of 0.662 MeV, u is ~ 0.07 cm^2/g, and relatively independent of material Z.

DATA ANALYSIS

There are some hidden features to Eq. (1) which have not been universally recognized in some of the previous density/thermal expansion literature. First, the refinement of Eq. (1) to include the dead time of the counting system has been ignored in some earlier works. The actual (corrected) count rate I differs from the measured (uncorrected) count rate R by the equation [9]

$$I = \frac{R}{1-R\tau} \quad , \tag{2}$$

where τ is the dead time of the counting system. For a given apparatus (fixed τ), the difference between I and R depends on the rate R, and will become negligible for small R.

Secondly, one should recognize that I_o, the (corrected) count rate in the absence of the sample material, can itself be represented by an equation of the form Eq. (1), where $(u \rho \ell)$ is replaced by $\Sigma\ u_i\ \rho_i\ \ell_i$, and the subscripts denote the materials in the path of the gamma beam (furnace walls, vacuum chamber walls, crucible walls, etc.). It is then clear that I_o will be a function of temperature, since $\rho_i \ell_i$ will vary as the temperature of each material changes. In addition, temperature can be expected to vary the size of collimation apertures as well, causing an additional dependence of I_o on T. These effects are taken into account by measurement of $R_o(T)$ as a function of temperature.

Finally, one recognizes that the sample dimension ℓ along the beam direction depends on temperature through the expansion of the containment crucible. These refinements lead to a modified form of Eq. (1) given by,

$$I(T) = \frac{R(T)}{1-R(T)\tau} = \frac{R_o(T)}{1-R_o(T)\tau} \exp\{-u\rho(T)\ell_1 \left[1 + \alpha(T-T_1)\right]\} \quad . \tag{3}$$

Here, R(T) and $R_o(T)$ denote measured (uncorrected) count rates at a sample temperature T, ℓ_1 is the sample length at T_1, and α is the mean linear expansion coefficient of the crucible over the range T_1 to T.

The mass attenuation coefficient is independent of the physical state of the material, and therefore is independent of temperature. However, since u results from both absorption and scattering processes, its value may be slightly dependent on the measurement geometry. Because accurate values of u may be lacking for a given material, it is advisable to measure u experimentally for each material investigated. This can usually be done at room temperature by measuring I through a length L of the material; then u is calculated from

$$u = \frac{1}{\rho L} \ell n\ (I_o/I) \quad . \tag{4}$$

The uncertainty in u results from the uncertainty in ρ and L for the material, since, in principle, one can count indefinitely at room temperature and thereby determine I_o and I to any precision.

An experimental determination of the counting system dead time τ is necessary since τ depends on the instruments and their particular settings. The two-source method was used to determine τ. Count rates were measured for irradiation by each of two sources separately, R_1 and R_2, and for both sources simultaneously, R_{12}. Using Eq. (2) and the restriction $I_1 + I_2 = I_{12}$ yields,

$$\tau = R_{12}^{-1} \left\{ 1 - \left[1 - \frac{R_{12}}{R_1 R_2} \left(R_1 + R_2 - R_{12} \right) \right]^{\frac{1}{2}} \right\} . \qquad (5)$$

The apparatus described here has $\tau \sim 3$ µsec; thus, for $R \sim 40,000$ cps, I is approximately 12% larger than R.

The gamma attenuation technique can be used to give both absolute and relative liquid density measurements. First, a form is presented for an absolute density determination. Solving Eq. (3) for $\rho(T)$ yields

$$\rho(T) = \frac{\ln \left[I_o(T)/I(T) \right]}{u \, \ell_1 \left[1 + \alpha(T - T_1) \right]} , \qquad (6)$$

where it is understood that I and I_o are determined from measured rates R and R_o using Eq. (2). To employ Eq. (6), u is first determined from Eq. (4) by measurements of I_o, and L and I of the sample material at a temperature (solid) where ρ is determined independently. The length ℓ_1 and expansion α of the crucible are also required. Next, $I_o(T)$ is measured, and finally, $I(T)$ is measured with the molten sample in the crucible to give $\rho(T)$ by Eq. (6). A number of independent measurements are thus required for this method of analysis.

Another approach, which may offer some advantages, is found by solving Eq. (6) for $(u \, \ell_1)$ at T_1 and equating this expression to the one for another temperature T_2. This leads to

$$\rho(T_2) = \frac{\rho(T_1)}{\left[1 + \alpha(T_2 - T_1) \right]} \frac{\ln \left[I(T_2)/I_o(T_2) \right]}{\ln \left[I(T_1)/I_o(T_1) \right]} . \qquad (7)$$

In Eq. (7), u and the absolute crucible dimension ℓ_1 do not enter the data analysis; instead, this approach requires independent knowledge of $\rho(T_1)$ in the melt from another method, and is therefore a relative density method.

A third analytical approach is found by writing Eq. (3) at a temperature T_1 and dividing throughout by Eq. (3) at T_2. One then has

$$\rho(T_2) \, \ell(T_2) - \rho(T_1) \, \ell(T_1) = \frac{\ell n \left[I(T_1) \, I_o(T_2)/I(T_2) \, I_o(T_1) \right]}{u} \cdot (8)$$

Unlike the previous two equations, this has a term $I_o(T_2)/I_o(T_1)$; if an experimentally-determined expression relating I_o at T_1 and T_2 can be found, some cancellation of terms may result. For I_o weakly-dependent on T, one can expect a functional dependence of the form $I_o(T_2)/I_o(T_1) = f(T_2 - T_1)$. Eq. (8) can then be rewritten as

$$\rho(T_2) = \frac{\rho(T_1)}{1 + \alpha(T_2 - T_1)} + \frac{\ell n \left[I(T_1) f / I(T_2) \right]}{u \, \ell(T_1) \, [1 + \alpha(T_2 - T_1)]} \cdot \qquad (9)$$

While this form requires knowledge of $\rho(T_1)$, u, and ℓ, it does not explicitly contain I_o. This may be particularly advantageous if f, the temperature dependence of I_o, is known, but I_o is not. This can readily occur for experimental arrangements where the furnace, controlled atmosphere chamber, and collimators are disturbed during sample loading. In addition, the elimination of I_o in Eq. (9) makes the determination of $\rho(T_2)$ much less sensitive to uncertainties in τ, since $I(T_1)$ and $I(T_2)$ will usually be less disparate than I(T) and $I_o(T)$.

Each of these methods requires some independently-determined parameter for the material, either ρ at a reference temperature or u. The accuracy to which $\rho(T_2)$ is determined may thus be limited by the error in $\rho(T_1)$ or u. It can be seen, therefore, that the gamma technique may be more useful in determining changes in density rather than the absolute densities themselves. Either Eqs. (6) or (7) above yields a fractional density change

$$\frac{\rho(T_2) - \rho(T_1)}{\rho(T_1)} = \frac{\ell n \left[I(T_2)/I_o(T_2) \right]}{[1 + \alpha(T_2 - T_1)] \, \ell n \left[I(T_1)/I_o(T_1) \right]} - 1 \cdot \qquad (10)$$

Of course, this expression is independent of $\ell(T_1)$, u, and an explicit knowledge of $\rho(T)$. Dividing Eq. (10) by $(T_2 - T_1)$ gives the mean volumetric thermal expansion coefficient α_v over the range T_1 to T_2. The precision of this value will depend on the count rate precision and a precise knowledge of $(T_2 - T_1)$.

PRECISION OF THE TECHNIQUE

By way of introduction to this section, the results of a pre-liminary experimental investigation of the precision of the method are briefly mentioned. Small dimensional steps in a precision-machined tool steel block were measured at ambient temperature using the technique. Since the gamma attenuation is a function of (ρl), the precision of measurement of length changes at constant ρ can be used to estimate the precision of measurement of small density changes. The step sizes in the steel block as determined by the gamma method agreed with metrology measurements to within the combined experimental error of both measurements. From this test, the sensitivity of the gamma method was confirmed to be $\Delta l/l \sim 10^{-4}$ cm/cm.

Analytical expressions for the precision of various parameters have been obtained by writing the rms deviation due to the propaga-tion of random errors through the data analysis expressions. If $\sigma(X_i)$ denotes the standard deviation in the parameter X_i, then one has the variance

$$\sigma^2(G) = \sum_i \left(\frac{\partial G}{\partial X_i}\right)^2 \sigma^2(X_i) \tag{11}$$

for the case where G is a function of the independent parameters X_i.

The precision analysis (using Eq. (11)) for τ is illustrated with a numerical example. For R_1 = 4000 cps, R_2 = 40,000 cps, R_{12} = 43,100 cps, and a total count time of 18,000 sec (5 hr), τ = 3.00 μsec $\pm$ 0.3%. Typically, then, one can achieve τ to better than 0.5%.

The determination of the mass attenuation coefficient u was also analyzed in terms of precision, using Eq. (11). As an example, ρ = 7.2 g/cm^3 $\pm$ .05%, l = 2 cm $\pm$ .05%, R and R_o $\pm$ .05%, and τ $\pm$ 1% together give $\bar{u}$ = .079 cm^2/g $\pm$ .11%. The uncertainty in calculated parameters is usually less sensitive to τ than to other parameters.

For typical values of R $\sim$ 5000-40,000 cps, one can readily achieve a count rate precision of 0.02% by counting for 5000-625 sec. Greater precision requires exceedingly long counting times, so that .02% will be used as an experimentally reasonable count rate pre-cision. The various analytical methods described above are now analyzed in terms of their precision.

The absolute density determination defined by Eq. (6) yields a fractional variance of $\rho(T)$ given by

$$\left(\frac{\sigma(\rho)}{\rho}\right)^2 = \left[u\,\rho\,\ell(T)\right]^{-2} \left\{ \left(\frac{\sigma(R_o)}{R_o}\right)^2 \frac{1}{(1-R_o\tau)^2} + \left(\frac{\sigma(R)}{R}\right)^2 \frac{1}{(1-R\tau)^2} \right.$$

$$\left. + \left(\frac{\sigma(\tau)}{\tau}\right)^2 \left[\frac{R_o\tau}{1-R_o\tau} - \frac{R\tau}{1-R\tau}\right]^2 \right\} + \left(\frac{\sigma(u)}{u}\right)^2 + \left(\frac{\sigma(\ell_1)}{\ell_1}\right)^2$$

$$+ (1 + \alpha\,\Delta T)^{-2} \left[(\Delta T)^2\,\sigma^2(\alpha) + \alpha^2\,\sigma^2(T) + \alpha^2\,\sigma^2(T_1)\right], \quad (12)$$

where $\ell(T) = \ell_1(1 + \alpha\,\Delta T)$, $\Delta T = T-T_1$, and $\ell_1 = \ell(T_1)$.

For comparison of the methods, a standard numerical example is chosen:

R = 16000 cps $\pm$ 0.02% R_o = 40,000 cps $\pm$ 0.02%

τ = 3 μsec $\pm$ 1% u = 0.07 cm^2/g $\pm$ 0.1%

ℓ_1 = 2 cm $\pm$ 0.05% T_1 = 300 K $\pm$ 1%

T = 700 K $\pm$ 1% α = 7 x 10^{-6}/K $\pm$ 10%

These values in Eqs. (6) and (12) yield $\rho(700$ K$)$ = 7.087 g/cm^3 $\pm$ 0.15%, where it is noted that u, τ, and ℓ_1 contribute most significantly to $\sigma(\rho)$. The variance contribution from the parameters is summarized in Table I, along with effects of changing some parameters.

The second method described for calculating the density is Eq. (7). While not including the analytical expression for $\sigma(\rho)$ which corresponds to Eq. (12), it should be noted that the sensitivity to R, R_o, τ, α, and T will be similar, and that $\sigma(\rho_1)/\rho_1$ (at T_1) will contribute directly to $\sigma(\rho_2)/\rho_2$ (at T_2), as do u and ℓ_1 in Eq. (12). To employ Eq. (7), an independently-measured value of ρ in the melt is needed. Crawley [11] has reviewed several methods for liquid metal density measurements; it appears that 0.2% may be a typical error for other methods. Using the above numerical example and $\rho(300$ K$)$ = 7.30 g/cm^3 $\pm$ 0.2% in Eq. (7) gives $\rho(700$ K$)$ = 7.018 g/cm^3 $\pm$ 0.21%. The same result obtains for an uncertainty in τ of $\pm$ 10%; indeed, this method is extremely insensitive to the imprecision of τ. It is, as expected, very sensitive to the uncertainty in $\rho(T_1)$. The accuracy of a value for $\rho(T_1)$ is the limiting factor in the resultant error of $\rho(T_2)$. Table II summarizes these results.

The third calculational method for determining ρ at a temperature T_2 was given by Eq. (9). Again, the fractional standard deviation $\sigma(\rho_2)/\rho_2$ was calculated from the contributions to the variance $\sigma^2(X_i)$ from each parameter. Inserting the standard set of values leads to $\rho(700\ K)$ with an uncertainty 0.23%. The relation

$$f = 1 + (10^{-5})\ (T_2 - T_1) \tag{13}$$

was used in Eq. (9) for this calculation, so that $I_0(T)$ increases 0.1% per 100 K. The primary contributions to the uncertainty in $\rho(T_2)$ using Eq. (9) are from $\rho(T_1)$ ($\pm$ 0.2%) and f ($\pm$ 0.1%). (This uncertainty in f allows that the value 10^{-5} in Eq. (13) may be in error as much as 25%.) As expected, the sensitivity to τ is small; in fact, an error in τ of as much as 50% had little effect on $\rho(T_2)$. It was found that $\rho(T_2)$, using Eq. (9), is particularly insensitive to u and ℓ_1. This is because u and ℓ_1 enter into the second term in Eq. (9), which is basically the correction added to $\rho(T_1)$ due to the change in temperature to T_2. For the above result, errors in ℓ_1 and u of 2% had little effect on the result. Table III summarizes these findings.

TABLE I. Method 1: $\rho(T)$ calculated from Eq. (6). Resultant % error in ρ shown for each of four sets of uncertainties in the input data parameters. Sets B, C, and D show the result of changing the % standard deviation from the values in set A. The % standard deviation for a parameter X_i is defined by 100 x $(\sigma(X_i)/X_i)$, where $\sigma(X_i)$ is the standard deviation of the parameter X_i.

| | | Set A | | % Standard Deviation | | |
| | | % Standard Deviation | Variance Contrib. (10^{-6}) | Set B | Set C | Set D |
Parameter						
$R(T)$	16000 cps	0.02	.045		0.2	
$R_0(T)$	40000 cps	0.02	.052		0.2	
τ	3 μsec	1.0	.75	10.0		
u	0.07 cm^2/g	0.1	1.00			0.4
$\ell(T_1)$	2.0 cm	0.05	.25			
T_1	300 K	1.0	.0004			
T	700 K	1.0	.002			
α	$7 \times 10^{-6}/K$	10.0	.079			
$\rho(T)$	7.127 g/cm^3	$\pm$ 0.15%		$\pm 0.87\%$	$\pm 0.34\%$	$\pm 0.41\%$*

* Same results using $\ell(T_1)$ $\pm$ 0.4%.

TABLE II. Method 2: $\rho(T_2)$ calculated from Eq. (7). See Table I caption for additional notes.

| | | Set A | | % Standard Deviation | | |
| | | % Standard Deviation | Variance Contrib. (10^{-6}) | | | |
Parameter				Set B	Set C	Set D
$\rho(T_1)$	7.30 g/cm^3	0.2	4.0	0.5		
$R(T_1)$	15400 cps	0.02	.041			0.2
$R_o(T_1)$	40000 cps	0.02	.048			0.2
$R(T_2)$	16000 cps	0.02	.044			0.2
$R_o(T_2)$	40100 cps	0.02	.052			0.2
τ	3 μsec	1.0	.0002		10.0	
T_1	300 K	1.0	.00002			
T_2	700 K	1.0	.0001			
α	7×10^{-6}/K	10.0	.078			
$\rho(T_2)$	7.018 g/cm^3	$\pm$ 0.21%		$\pm 0.50\%$	$\pm 0.21\%$	$\pm 0.48\%$

In each of the three methods described, resultant error due to uncertainties in T and α was very small. Similarly, the values of count rate intensities $\pm$ 0.02% were sufficiently precise to cause negligible error in the calculated density. Increasing the rate uncertainties to 0.2%, which represents a 100X reduction in counting time, approximately doubles the resultant uncertainty in the density, for the examples given.

In calculating the mean volumetric expansion (using Eq. (10)) and its uncertainty, R_o and R were chosen as above ($\pm$ 0.02%), with values such that $\alpha_v \sim 10^{-4}$/K. Values for τ and α were unchanged. Using a precision for T of $\pm$ 0.2%, the resultant uncertainty in α_v was $\pm$ 2.3%, $\pm$ 4.5%, and $\pm$ 8.6% for $\Delta T = (T_2-T_1)$ of 200 K, 100 K, and 50 K, respectively. Systematic errors in temperature contribute significantly less to the analysis.

SUMMARY

An apparatus for the measurement of liquid densities at high temperatures using the gamma attenuation method was described, along with an electronic detection system with long-term stability against drift and fluctuations. Experimental and theoretical considerations for choice of sample size and gamma source were also presented.

TABLE III. Method 3: $\rho(T_2)$ calculated from Eq. (9). See Table I caption for additional notes.

| Parameter | | Set A | | % Standard Deviation | | |
		% Standard Deviation	Variance Contrib. (10^{-6})	Set B	Set C	Set D
$\rho(T_1)$	7.30 g/cm³	0.2	4.3		0.5	
$R(T_1)$	15400 cps	0.02	.045			
$R(T_2)$	16000 cps	0.02	.045			
τ	3 μsec	1.0	.0004	50.0		
u	0.07 cm²/g	0.1	.001	2.0		
$\ell(T_1)$	2.0 cm	0.05	.0003	2.0		
f	1.004	0.1	1.0			0.4
T_1	300 K	1.0	.0004			
T_2	700 K	1.0	.002			
α	7 x 10⁻⁶/K	10.0	.078			
$\rho(T_2)$	7.022 g/cm³	± 0.23%		+0.28%	±0.53%	±0.46%

Three data analysis methods for determining the liquid density were presented, along with a precision analysis for each. Depending on the experimental conditions, the methods offer the ability to use different sets of input data, which can be advantageous where experimental determination of some of the parameters is irreproducible or imprecise. In each of the data analysis methods, the necessary corrections for counting system dead time and crucible expansion have been included.

Method 1 allows determination of the liquid density $\rho(T)$ through Eq. (6), and requires an independent determination of the mass attenuation coefficient for the material. Imprecision in measurement of this coefficient, the system dead time τ, and the crucible dimension ℓ_1 are the most significant contributors to the resultant imprecision in $\rho(T)$.

The second method to calculate $\rho(T)$ uses Eq. (7), and requires independent knowledge of $\rho(T_1)$, the liquid density at some fiducial temperature. However, it does not require precise knowledge of the crucible dimension. Using this method, the resultant imprecision of $\rho(T)$ is primarily due to the error in $\rho(T_1)$, and relatively insensitive to measurement of the dead time τ.

Method 3 (Eq. (9)) is advantageous for data analysis when specific values for I_o, the gamma intensity through the experimental system in the absence of the sample, are not known. As in method 2, a value is needed for the liquid density at a single reference temperature; the error in this value is the primary contributor to the resultant imprecision. Uncertainties in the crucible dimension, dead time, and mass attenuation coefficient are relatively unimportant in this data analysis method.

Use of experimentally reasonable values for the input parameters and their precisions yields a similar resultant density precision from the three methods, on the order of 0.2%.

REFERENCES

1. I. G. Dillon, F. E. LeVert, P. A. Loretan, G. U. Menon, F. M. Siddiqi, and H. J. Tarng, Nucl. Tech. 12, 307-313 (1971).

2. F. E. LaVert, I. G. Dillon, and H. J. Tarng, Rev. Sci. Instrum. 44, 313-315 (1973).

3. G. Döge, Z. Naturforschg. 21a, 266-269 (1966).

4. A. S. Basin and A. N. Solov'ev, Zhur. Priklad. Mekh. Tekhn. Fiziki 6, 83-87 (1967).

5. V. I. Yavoyskiy et al., Russ. Metall. No. 4, 44-47 (1974).

6. W. H. Gardner, G. S. Campbell, and C. Calissendorff, Soil Sci. Soc. Amer. Proc. 36, 393-398 (1972).

7. K. Preiss, Soil Sci. 110, 151-156 (1970).

8. E. Elias, Y. Segal and A. Notea, Nucl. Tech. 21, 57-66 (1974).

9. R. D. Evans, The Atomic Nucleus, McGraw-Hill, New York, 1955.

10. A. Picot, Rev. Sci. Instrum. 47, 385-386 (1976).

11. A. F. Crawley, Inter. Metall. Rev. 19, 32-48 (1974).

DENSITY AND THERMAL EXPANSION IN LIQUID METALS AND ALLOYS

P.J. Desré
Laboratoire de Thermodynamique et Physico-
Chimie Métallurgiques associé au CNRS
Domaine Universitaire, Saint Martin d'Hères, France

L.D. Lucas
I.R.S.I.D. - Station d'Essais
57210 Maizières les Metz, France

Results of density measurements in pure liquid metals will be briefly presented and discussed. A classification based on an empirical relation between density and thermal expansion is proposed.

Theoretical approach of the thermal expansion in liquid metals will be discussed in relation to the structure. The use of the Virial equation of state in evaluating the thermal expansion coefficient is not easy because of the difficulty of finding a good effective pair potential specific to the liquid metallic state. Preference is given to the compressibility equation of state which permits expression of the thermal expansion against isothermal compressibility and the dependance with temperature of the pair correlation function.

The thermal expansion coefficient in liquid alloys will be discussed in terms of the variation with temperature of the volume of mixing. For some alloys abnormal behaviour in the variation of the volume of mixing with temperature will be related to the evolution of the alloy structure.

ANOMALOUS THERMAL EXPANSION OF SODIUM CHLORATE

T.N. Wathore, Dept. of Physics
Marathwade University
Aurangabad, Maharashtra, India

R.G. Kulkarni, Dept. of Physics
Government Science College
Aurangabad, India

ABSTRACT

The lattice parameter a_t of Sodium chlorate has been measured in the temperature range 32 to 249°C using high temperature X-ray diffractometer and a 15 cm diameter symmetrically focusing back-reflection camera. a_t in A° is related to the temperature t in °C by:

$$a_t = 6.5697 + 247.5451 \times 10^{-6} t - 0.6350 \times 10^{-8} t^2$$
$$+ 12.8099 \times 10^{-10} t^3$$

The linear thermal expansion coefficient α is calculated by differentiation of a_t with respect to temperature and the values of α above 130°C exhibit anomalous behaviour. The analysis of the anomalous expansion data gives a thermal defect formation energy of 1.41 $\pm$ 0.05 ev.

1. INTRODUCTION

Thermal expansion of a crystal lattice occurs because of anharmoic contributions to the crystal potential. Any point defect in the crystall will increase the anharmoicity of the potential, which will give rise to a rapid increase in thermal expansion (α) at high temperatures i.e. an anomalous behaviour for α. According to Lidiard (1957) such an anomalous expansion is expected for ionic crystals in the intrinsic range of temperatures. Lawson (1950), who

studied the thermal expansions of silver halides, ascribes such an
anomalous behaviour of α to the thermal generation of point defects.

The thermal expansion of sodium chlorate ($NaClO_3$) has been
measured by Sharma (1950) and Deshpande et al (1960) using interferometric and lattice parametric methods respectively. Their
results are not only in disagreement with each other at higher
temperatures but also exhibit anomalous behaviour for α. In order
to remove this amibiguity and to throw some light on the anomalous
behaviour of α, we present the results of an experiment in which
the lattice parameter of $NaClO_3$ was measured with precision using
both X-ray powder-photographic and X-ray diffractometric techniques
in the temperature range 32 to 249°C. The anomalous expansion data
was used in obtaining information about thermally produced point
defects and also in calculating the defect formation energies.

2. EXPERIMENTAL PROCEDURE

The powder samples of $NaClO_3$ with a specified purity of 99.99%
were obtained from Riedel Dettaen Agseelze, Hanover, Germany. To
obtain uniform particle size the powders were filtered through a
44 μm sieve.

Measurements were carried out with a diffractometer and back-
reflection focusing camera. The high temperature X-ray diffractometric studies were conducted at Bhabha Atomic Research Centre,
Trombay. The experimental setup of the diffractometer has been described elsewhere (Momin et al 1971). The high temperature X-ray
diffraction patterns with ten reflections were scanned at 1/2º 2θ
min^{-1} in the temperature range 32º to 232ºC in air. Both the X-ray
diffractometric and powder photographic studies were made using
filtered CuK_α radiation. A symmetrically focusing back-reflection
camera of 15 cm diameter was also used for obtaining powder photographs at eleven temperatures between 32 to 249°C.

3. RESULTS AND DISCCUSION

3.1 Lattice Parameter and Thermal Expansion

The lattice parameter at different temperatures was determined
accurately from the diffractometric data and powder photographic data
using Nelson and Riley (1945) and $\phi\tan\phi$ error functions respectively.
The lattice parameter obtained for each temperature is listed in
Table 1. The standard errors calculated by the methods of Jette and
Foote (1935) are also tabulated. The lattice parameter found in the
present work agrees favourably with the values of Deshpande et al
(1960) up to 120°C and at higher temperatures it varies from

Deshpande's values by 7-8%. The variation of lattice parameter
with the temperature (column 4 of Table 1) is non-linear and the
least squares fitting to these data points gives the following
equation:

$$a_t = 6.5697 + 247.5451 \times 10^{-6}t - 0.6350 \times 10^{-8}t^2 + 12.809 \times 10^{-10}t^3$$

$$\ldots\ldots\ldots(1)$$

where a and t are expressed in $\overset{o}{A}$ and ^oC respectively. The corresponding equation for the coefficient of thermal expansion obtained
by differentiation of a_t with respect to temperature is

$$\alpha_t = 37.679 \times 10^{-6} - 0.1933 \times 10^{-8}t + 5.8495 \times 10^{-10}t^2$$

$$\ldots\ldots\ldots(2)$$

where α and t are expressed in $^oC^{-1}$ and ^oC respectively. The error
in α is estimated to be less than 2%.

<u>TABLE I</u>

Lattice Parameter of Sodium Chlorate at Various Temperatures

t(C)	X-ray photographic a(A^o)	X-ray diffra- ctometric a(A^o)	Lattice Parameter
32	6.5777 ± 0.0002	6.5775 ± 0.0002	6.5776
54	6.5834	--	6.5834
66	6.5867	--	6.5867
103	--	6.5964	6.5964
157	6.6132	--	6.6132
187	6.6242	--	6.6242
206	--	6.6316	6.6316
217	6.6364	--	6.6364
233	--	6.6436	6.6436
245	6.6488	--	6.6488
249	6.6505	--	6.6505

The variation of α_t with temperature of this work, of Sharma
(1950) and of Deshpande et al (1960) are presented in Table II. A
comparative study of the coefficients of expansion at different
temperatures given by previous workers (Sharma 1950; Deshpande et al
1960) and the present work shows that the values of Sharma (1950)
near room temperature are high than our values by about 20.5; whereas Deshpande's values are about 9% higher. The gradient of the α-t

curve of this work agrees reasonably with that of Despande et al
(1960). However, the present work gives a higher gradient than that
of Sharma (1950), so that at higher temperatures the microscopic
X-ray values are much larger than the macroscopic values of Sharma
(1950). The discrepancy observed between the α_t values of our investigation and of previous work (Sharma, 1950; Deshpande et al 1960)
may be explained in terms of crystal imperfections. According to
Lidiard (1957) the point defects in crystals at higher temperatures
contribute substantially to the thermal expansion of the crystal.
He says that if the intrinsic defects are of Schottky type, the thermal expansion coefficient obtained from macroscopic method should be
higher than that obtained from measurement of lattice parameter, and
if the α values obtained by both methods agree then the defects are
of Frenkel type. The α-values of the present work are smaller than
the macroscopic values of Sharma up to 180°C and later on our α
values tend to be higher than that of Sharma's. Similiar behaviour
was observed by Deshpande et al (1960). We, in agreement with Deshpande et al (1960) ascribe this behaviour of α to point defects.

TABLE II

Thermal Expansion Coefficient of Sodium Chlorate at Various Temperatures

$t(C)$	$\alpha(10^6 K^{-1})$ Present Work	$\alpha(10^6 K^{-1})$ Deshpane et al (1960)	$\alpha(10^6 K^{-1})$ Sharma (1950)
32	38.22	41.37	44.78
54	39.28	42.82	46.22
66	40.10	43.85	47.05
103	43.69	48.12	49.71
157	51.80	57.27	53.94
187	57.77	63.86	56.48
206	62.10	68.58	58.15
217	64.81	71.52	59.14
233	68.99	--	60.62
295	72.32	--	--
249	73.46	--	--

3.2 Defect Formation Energy

The thermal expansion data obtained in this work, by Sharma
(1950) and by Deshpande et al (1960) show an anomalous behaviour at
higher temperatures (above 130°C). Lidiard (1957) suggests that such
an anomalous expansion is expected for ionic crystals in the intrinsic
range of temperatures. Merriam et al (1962) analysed the anomalous
thermal expansion in NaCl and found a reasonable value for the formation energy of Schottky defects. However, Fischmeister (1956), found

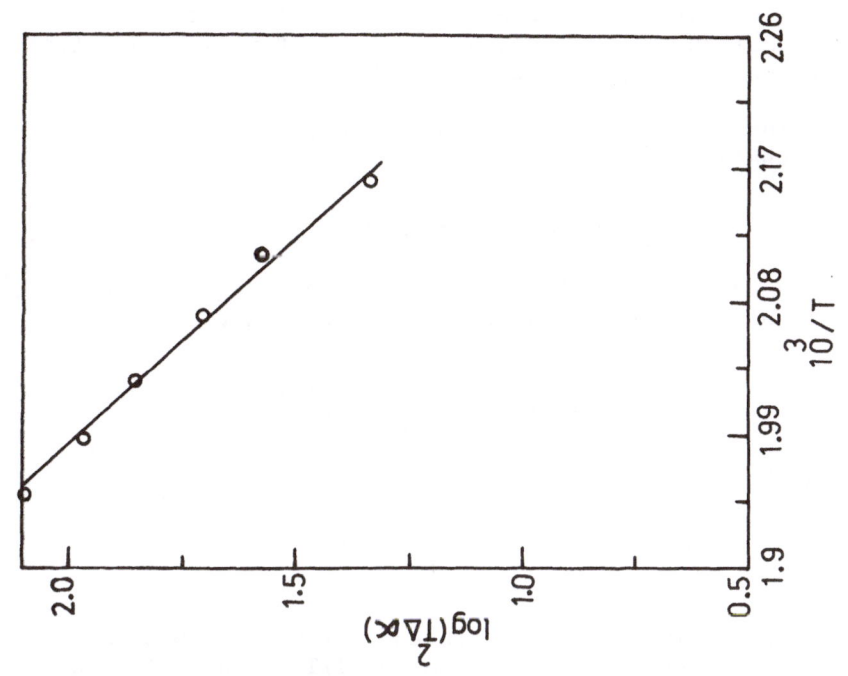

Fig 2: Determination of defect formation energy

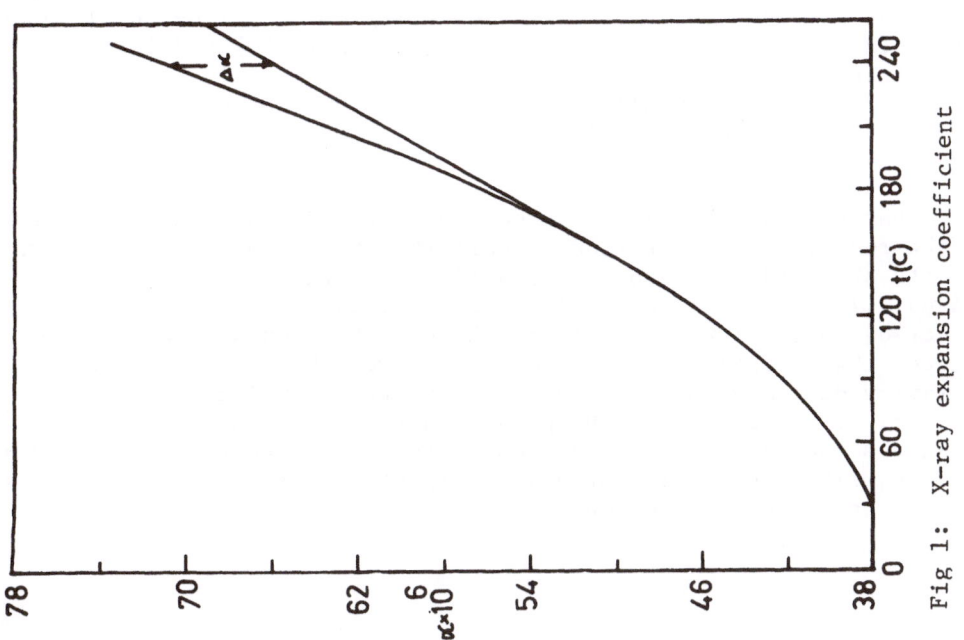

Fig 1: X-ray expansion coefficient

considerable disagreement between the activation energies obtained
from the anomalous thermal expansion and electrical conductivity
data in the case of many ionic crystals.

According to Lidiard (1957) the presence of defects produce a
fractional change in volume:

$$\delta = \frac{\Delta V}{V} = A\ e^{-E/2KT} \qquad \ldots\ldots(3)$$

where A is a constant and E is the formation energy of the defect.
The additional expansion α due to this change in volume is:

$$\Delta\alpha = \frac{1}{V}\ \frac{d}{dT}\ (\Delta V) = \frac{A}{2K^2}\ e^{-E/2KT} \qquad \ldots\ldots(4)$$

so that

$$T^2(\Delta\alpha) = \frac{A}{2K}\ e^{-E/2KT} \qquad \ldots\ldots(5)$$

The α_T values obtained from the present X-ray data shown in Figure 1,
were used to plot graphs of $\ln(T^2\ \Delta\alpha)$ vs $1/T$ (Figure 2). The graph
(Figure 2) exhibits essentially straight line behaviour, and so the
least-squares fitting was used to calculate the slope. The defect
formation energy derived from this slope is 1.41 + 0.05 ev. Using
a similar procedure, defect formation energies of 1.2 ev and 1.35 ev
were obtained from macroscopic data of Sharma (1950) and lattice
parametric data of Deshpande et al (1960) respectively. All the
three values are far smaller than the formation energy of 2.18 ev
obtained from the electrical conductivity measurements of Ramsastry
et al (1971).

Sastry et al (1970) have shown that the long wavelength region
of the fundamental absorption is perturbed by the presence of thermal
defects above 130°C and obtain a defect formation energy of value
1.4 ev. This value agrees well with our value. There is some con-
troversey in the literature as to whether there should be agreement
between the formation energies obtained from the electrical conductivity data and those calculated from the anomalous thermal expansion. It appears that the present-day understanding of the
anomalous thermal expansion of ionic crystals at temperatures approch-
ing the melting-point is not quite complete and that point defects do
affect the thermal expansion.

<u>REFERENCES</u>

(1)   Deshpande, V.T. and Mudholkar, V.M. 1960 Acta Cryst. <u>13</u>, 483.

(2)   Fischmeister, H.F. 1956 Acta-Cryst. <u>9</u>, 416.

(3)   Jette, E.R. and Foote, F. 1935, J. Chem. Phys. <u>3</u>.

(4)   Lidiard, A.B., 1957 Handb, Phys. <u>20</u>, 246.

(5)   Merriam, M.F. Smolwchowski, R., and Wiegand, D.A.,1962,
      Phys. Rev. <u>125</u>, 65.

(6)   Momin, A.C., Mathews, M.D. and Karkhanvala, M.D. 1971,
      Indian J. Chem. 9 582.

(7)   Nelson, J.B. and Riley, D.P. 1945, Proc. Phys. Soc., <u>57</u>, 477.

(8)   Sastry, S.B.S., Tripathi, R.B., and Ramasastry, 1970, J. Phys.
      Chem. Solids, <u>31</u>, 2765.

(9)   Sharma, S.S. 1950, Proc. Indian Acad. Sci. <u>A31</u>, 83.

(10) Ramasastry, C., Viswanatha Reddy, K., and Murthy, V.S., 1971,
      Proc. R. Soc. Lond, <u>A.325</u>, 347.

LATTICE THERMAL EXPANSION AND LATTICE VIBRATIONS OF LEAD NITRATE

G.K. Bichile* and R.G. Kulkarni

Department of Physics, Marathwada University

*Department of Physics, P.G. Centre College, Jalna

ABSTRACT

The lattice parameter a_T of lead nitrate has been measured in the temperature range 263 to 621 K using a high temperature X-ray diffractometer and a 15 cm diameter symmetrically focusing back reflection camera. a_T is found to increase parabolically with temperature. The linear thermal expansion coefficient α is calculated by differentiation of a_T with respect to temperature and it varies linearly with temperature. The temperature dependence of the lattice vibrational frequencies is found using the present thermal expansivity values and the Gruneisen parameter γ obtained at room temperature. It is found that the lattice vibrational frequencies of lead nitrate decrease on the average (geometric mean) by 6.35% when the powder sample is heated from 298 to 621 K.

1. INTRODUCTION

The thermal expansion of a solid body arises because of the anharmonicity of the potential energy of a crystal and is directly related to the motion of the atoms in the interatomic potential of the crystal. The vibrational frequencies of the atoms in the crystal lattice are volume dependent. The volume dependence of the frequency is defined in terms of the Gruneisen parameter γ, which is a measure of the anharmonicity of the interatomic potential. Thus the measurement of thermal expansion not only helps in determining γ, but also, in finding the temperature dependence of the lattice vibrational frequencies (Schaur 1964, 1965).

In the present work, the thermal expansion coefficient of lead nitrate [Pb(NO$_3$)$_2$] has been found at various temperatures by the x-ray method with the object of extending the existing low temperature measurements to higher temperatures and to determine the temperature dependence of the lattice vibrational frequencies.

2. EXPERIMENTAL

The powder samples of Pb(NO$_3$)$_2$ were obtained from British Drug House, Ltd., London and had a specified purity of 99.99%. This powder sample was filtered through a 44 μm sieve to give uniform particle size.

The high temperature X-ray diffractometer studies were conducted at Bhabha Atomic Research Centre, Trombay. The experimental set up used to obtain the diffraction pattern has been described elsewhere (Momin et al 1971). All X-ray diffraction studies were made using filtered CuK$_\alpha$ radiation (36 kV 18 mA). A symmetrically focusing back-reflection camera of 15 cm diameter was also used for obtaining powder photographs at elevated temperatures with filtered copper radiation.

3. RESULTS

The accurate determination of the lattice parameter from the diffractometric data and powder photographic data was done using the Nelson and Riley (1945) and ϕtanϕ error functions respectively. The lattice parameter a$_T$ was measured in the temperature range 263 to 621 K and a$_T$ varies non-linearly with temperature. The least squares fitting to the a$_T$ data points gives the following equation:

$$a_T = 7.801 + 1.5929 \times 10^{-4}T + 11.7023 \times 10^{-8}T^2 \qquad \ldots\ldots\ldots(1)$$

where 'a' and 'T' are expressed in $\overset{o}{A}$ and K respectively.

The thermal expansion coefficient (α) obtained by differentiation of a$_T$ [eqn.(1)] with respect to temperature is:

$$\alpha_T = 20.69 \times 10^{-6} + 2.48 \times 10^{-8}T \qquad \ldots\ldots\ldots(2)$$

where α and T are expressed in K^{-1} and K respectively. The error in α is estimated to be less than 2%. The present results (equation 2) suggest that α varies linearly with temperature in accordance with Srinivasan's (1955) low temperature (348-488 K) results.

4. DISCUSSIONS OF RESULTS

The temperature dependence of the lattice vibrations was determined by using the procedure of Kulkarni et al. (1977). The temperature variation in lattice vibrational frequencies is

$$\frac{\nu_g(T)}{\nu_g(T)} = \Pi_i \left(\frac{\nu_i(T)}{\nu_i(To)} \right)^{1/3N} = 1 - \int_{T_o}^{T} \gamma \ \beta \ dT \qquad \ldots\ldots(3)$$

where $\nu_i(T)$ is the frequency of the ith vibration at temperature T, $\nu_g(T)$ is the geometric mean, β is the volume expansivity, γ is the Grüneisen parameter, T is the temperature, T_o is the reference temperature (298 K), and 3N is the number of the vibrational modes in the crystal of N atoms.

The variation in $\nu_g(T)/\nu_g(To)$ with temperature was calculated using equation (3) assuming a constant value of $\gamma = 1.45$ above 273 K (Bichile, 1976) and $\beta = 3\alpha$ (equation 2). The results are shown in Table 1. It is found that the average of the average (geometric mean) of the lattice vibrational frequencies of lead nitrate decrease by 6.35% when the powder sample is heated from 298 to 621 K.

TABLE 1

Mean variation of the lattice vibrational frequencies of $Pb(NO_3)_2$ with temperature

Temperature (K)	$1 - \int_{T_o}^{T} \gamma\beta \ dT$
318	0.997
338	0.994
358	0.991
378	0.988
398	0.985
418	0.982
438	0.979
458	0.976
478	0.973
498	0.970
518	0.967
538	0.963
558	0.961
578	0.958
598	0.955
618	0.952
638	0.949

$\gamma = 1.45 \qquad \beta = 3\alpha = 3(20.69 \times 10^6 + 2.84 \times 10^{-8}T)$

$T_o = 298$ K.

References

Bichile, G.K. (1976) Ph.D. thesis, Marathwada Univerisity.

Kulkarni, R.G., Bichile, G.K., Kharkhanavala, M.D.
and Momin, A.C. (1977) J. Phys. Soc., Japan 42 971.

Momin, A.C., Mathews, M.D., and Kharkhanvala, M.D. (1971),
Indian J. Chem. 9 582.

Nelson, J.B. and Riley, D.P. (1945), Proc. Phys. Soc. 57 477.

Schaur, A. (1964), Can. J. Phys. 42, 1857.

Schaur, A. (1965), Can. J. Phys. 43, 523.

Srinivasan, R. (1955), Proc. Ind. Acad. Sci. A41, 49.

EXPANSION MEASUREMENT USING SHORT CYLINDRICAL

SEAL: THEORY AND EXPERIMENT

S. T. Gulati and H. E. Hagy

Research and Development Laboratories
Corning Glass Works
Corning, New York 14830

ABSTRACT

In this paper we examine the effect of the seal length on the magnitude of residual stresses and the expansion differential derived from them. Two potential errors can arise which, if uncorrected, can lead to substantially lower values of expansion differential. The first error arises due to the finite length of the seal which changes the three-dimensional stress distribution significantly, and makes the standard formulas based on infinitely long seal invalid. It is necessary to modify the standard formulas by introducing the shape factor which depends on the geometry and elastic properties of the seal members. The second error is associated with attributing the observed birefringence (tangential viewing) solely to the axial stress. This error, which is absent in the infinitely long cylindrical seal, can be eliminated by taking into account the contributions of

radial and circumferential stresses. The theoretical
solution is obtained via finite element analysis for
glass-to-glass, glass-to-Kovar* and glass-to-steel seals
of different aspect ratios. The experimental verification of the theory is accomplished by fabricating a
glass-to-glass seal and examining the changes in birefringence as the seal is shortened by the technique of
successive mill-grinding. The paper concludes with a
series of curves for determining the shape factor for a
given bead seal.

INTRODUCTION

The bead seal[1-4] has served a useful purpose over
the many years since its introduction. In glass-to-
metal seals, for example, it has provided an efficient
means for determining the expansion differential between
glass and metal and has led to the optimization of sealing glasses. This task is generally accomplished by
making the cylindrical seal with metal as the core material and a suitable glass as the cladding material, and
examining the residual stresses at the interface under
polarized light.[5] The seal is viewed either axially or
tangentially[6-8], Fig. 1, and the observed birefringence,
which is linearly related to the stresses and hence the
expansion differential, provides a direct measure of the
expansion differential between glass and metal at the
setting temperature of the glass.

The above approach assumes that the cylindrical

* Trademark of the Carborundum Company. Kovar is a 29%
  Nickel - 17% Cobalt - Iron Sealing Alloy conforming
  to ASTM Specification F15.

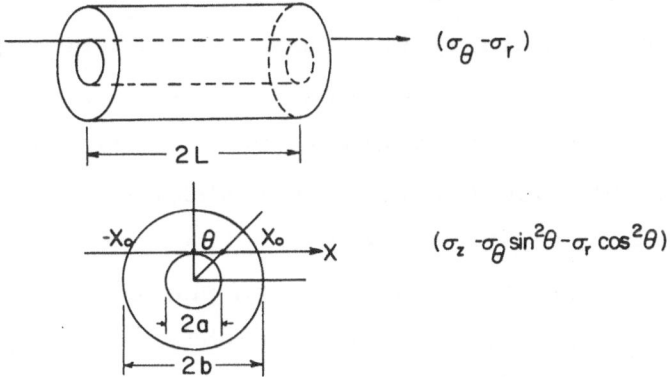

Fig. 1   Axial and Tangential Viewing of Bead Seal

seal is long compared to its diameter. The elastic
solution for residual stresses was first derived by
Portisky[4] using the general solution of Lame[9] and since
his final expressions for stresses contained an error,
it was rederived by Redston and Stanworth[2] and more
recently by Takagi[10] and by Martin[11] who allowed for the
viscous flow of glass and plastic yielding of the metal.
Takagi's solution is somewhat more exact as it does not
assume the equality of Poisson's ratios of the seal
members. Kishii[12,13] has used this solution for examining glass-to-Dumet seals.

When the seal aspect ratio (length/diameter) is
less than 10, the end effects become important and the
stresses vary along the length of the seal as well as
along the tangential line of sight. One obtains an integrated average value of optical retardation which is
smaller than that for the infinitely long seal. The
smaller the aspect ratio, smaller is the retardation.

Thus the expansion differential inferred from retardation would correspondingly get smaller with the aspect
ratio. This is physically absurd since the expansion
differential between two materials must remain constant
regardless of the aspect ratio. For short seals, therefore, it is necessary to correct for the aspect ratio by
introducing a shape factor similar to the recent approximate theory of the narrow sandwich seals.[14,15] The
elastic solution of the finite cylindrical seal is obtained rather efficiently by resorting to finite element
analysis and is compared with the exact closed-form solution of the infinite cylindrical seal. The two shape
factors, one for tangential viewing (K_t) and the other
for axial viewing (K_a), are simply the ratios of corresponding stresses for seals of finite and infinite length.

In addition to the aspect ratio, the shape factor
is also affected by the elastic properties of seal members and by the cross-sectional geometry of the seal.
These effects can readily be examined by finite element
analysis. In this paper we present the results for
glass-to-glass, glass-to-Kovar and glass-to-steel seals
with elastic moduli ratio ranging from 1 to 3 and glass/
metal diameter ratio ranging from 1.5 to 3. The shape
factor and hence the error, for a given aspect ratio, is
found to be more severe for large values of elastic
moduli ratio and diameter ratio. The finite element
predictions of the shape factor are verified experimentally by fabricating a glass-to-glass seal and examining
the changes in retardation as the aspect ratio is reduced

from 18 to 1 by the technique of successive mill-grinding. The agreement is found to be good.

THEORY

For the infinitely long seal the radial, circumferential and axial stresses in the glass cladding are given by

$$\sigma_r = A_2 \left(1 - \frac{b^2}{r^2}\right) \tag{1}$$

$$\sigma_\theta = A_2 \left(1 + \frac{b^2}{r^2}\right) \tag{2}$$

$$\sigma_z = \text{constant} \tag{3}$$

where r defines the radial position, r = b is the outer surface of the seal and A_2 is the constant of integration. When the seal is viewed tangentially, one of the stresses normal to the line of sight is σ_z and the other stress is

$$\sigma = \sigma_r \cos^2\theta + \sigma_\theta \sin^2\theta \tag{4}$$

which by virtue of eqns. 1 and 2 may be written as

$$\sigma = A_2 \left[1 - \left(\frac{b}{a}\right)^2 \frac{\cos 2\theta}{\sec^2\theta}\right] \tag{5}$$

Here a is the radius of the metal core. Let us define

$$x = a \tan\theta \tag{6}$$

and integrate σ along the line of sight from $-x_o$ to x_o.

$$\int_{-x_0}^{x_0} \sigma\,dx = A_2\,a \int_{-\beta}^{\beta} \left[\sec^2\theta - \left(\frac{b}{a}\right)^2 \cos 2\theta\right] d\theta = 0 \qquad (7)$$

Thus the integrated value of stress difference, $\sigma_z - \sigma$, along the line of sight is nothing but the average value of axial stress which we shall denote by

$$\int_{-x_0}^{x_0} (\sigma_z - \sigma)\,dx = \int_{-x_0}^{x_0} \sigma_z\,dx = \tilde{\sigma}_z \qquad (8)$$

Let us note that the vanishing of the integrated value of σ is a direct consequence of the form of the solution, eqns. 1 and 2, for stresses in infinitely long seals. In short seals the stress distribution deviates from that given by eqns. 1 through 3 and it is necessary to not only use average value of axial stress along the line of sight but also to allow for any non-zero contribution from σ.

The final expression relating the axial stress and optical retardation in an infinite seal is given by

$$\sigma_z^{\infty} = \frac{R_\infty}{C\ell} = \frac{(E_g/1-\nu_g)\,\delta_\infty}{\left[\frac{E_g}{E_m}\left(\frac{b}{a}\right)^2 - \frac{E_g}{E_m} + 1\right]} \qquad (9)$$

in which R_∞ is the optical retardation, C is the stress-optic coefficient of glass, E_g and E_m are elastic moduli of glass and metal respectively, ν_g is the Poisson's ratio for glass, δ_∞ is the expansion differential between glass and metal at the setting temperature of glass and ℓ is the path length which is given by

$$\ell = 2\sqrt{b^2 - a^2}$$

Thus for the infinite seal,

$$\delta_\infty = \gamma R_\infty / C \tag{10}$$

whereas for the short seal of half length L

$$\delta_L = \gamma R_L / C \tag{11}$$

In the above expressions γ is the seal constant involving seal geometry and elastic properties. Since $\delta_L < \delta_\infty$, the true value of δ is given by

$$\delta_{true} = \delta_L / K_t \tag{12}$$

where

$$K_t = R_L / R_\infty = \tilde{\sigma}_z / \sigma_z^\infty < 1 \tag{13}$$

In the case of axial viewing of long seal, the optical retardation is related to the difference of cirfumferential and radial stresses through

$$(\sigma_\theta - \sigma_r)^\infty = \frac{R_\infty}{C\ell} = \frac{2E_g \delta_\infty}{\Delta}(\tfrac{b}{a})^2 \left[1 + \nu_g + \lambda m(1 + \nu_m)\right] \tag{14}$$

where ν_m is the Poisson's ratio for the metal and

$$\ell = 2L , \qquad \lambda = (\tfrac{b}{a})^2 - 1 , \qquad m = E_g / E_m$$

$$\Delta = \lambda^2 m^2 (2\nu_m - 1)(\nu_m + 1) + \lambda m (4\nu_g \nu_m + \nu_m - \nu_g \lambda - \lambda - 3)$$
$$- (1 + \tfrac{b^2}{a^2} - 2\nu_g^2 + \nu_g \lambda)$$

The expressions analogous to eqns. (10) through (12) are

$$\delta_\infty = \eta R_\infty / C \tag{15}$$

$$\delta_L = \eta R_L / C \tag{16}$$

and

$$\delta_{true} = K_a \delta_L \tag{17}$$

where η is the seal constant and K_a is the second shape factor and is given by

$$K_a = \frac{R_\infty}{R_L} = (\sigma_\theta - \sigma_r)^\infty / (\sigma_\theta - \sigma_r)_L > 1 \tag{18}$$

Let us note that both K_t and K_a are functions of seal length, glass/metal diameter ratio and the elastic properties of seal members.

In general the axial and circumferential stresses in the glass cladding are of the same sign and the radial stress is of opposite sign. Thus if the axial stress is tensile, the seal geometry and glass properties should be chosen in such a way as to keep the circumferential stress below a safe allowable value. If, on the other hand, the axial stress is compressive then the radial stress will be tensile and should be measured and limited to a safe value to insure the mechanical integrity of the seal. Under some conditions, however, the setting point of the glass can be different for the radial - tangential stresses than for the axial stress.[11] This occurs, for example, when the expansion curves of the metal and glass cross near the strain point of the glass. In such cases, both the radial and axial stresses can be tensile and their magnitude should be checked out photo-elastically using above equations.

FINITE ELEMENT ANALYSIS

The finite element program used for the analysis of stresses in short cylindrical seals was ANSYS.[16] In view of the symmetry of the seal about its axis and the mid-length, it is necessary to examine only a quarter of the seal cross-section. The finite element mesh selected for the analysis is shown in Fig. 2. The axi-

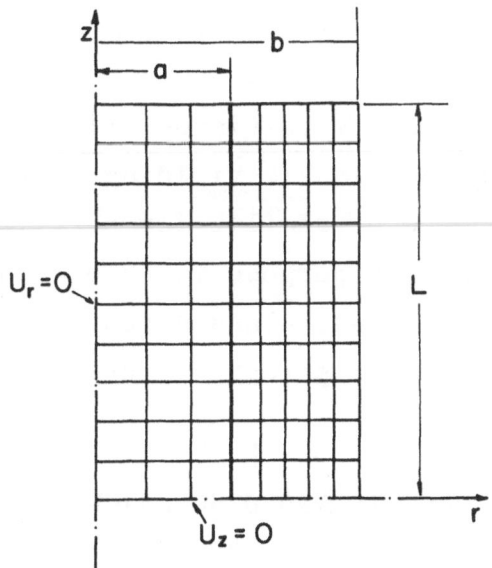

Fig. 2 Finite Element Mesh

symmetric isoparametric element was used. In view of the relative importance of stresses in glass, the glass section was divided into five layers of equal thicknesses and the metal section was divided into three layers.

The seal was divided into ten equal segments along its length. The mesh size is rather arbitrary and can be refined even further to obtain a more accurate picture of stresses. However, the computation cost can multiply rapidly making the economics unfavorable. The boundary conditions chosen for the displacements along the lines of symmetry were

$$U_z = 0 \qquad \text{along } z = 0$$

and

$$U_r = 0 \qquad \text{along } r = 0$$

The corrections due to viscous flow of the glass during annealing and plastic yielding of the metal, if any, were assumed to be similar for both long and short bead seals. Since our purpose is to evaluate the shape factors which are dimensionless stress ratios, the elastic analysis of thermal stresses provided by the ANSYS program is adequate.

RESULTS AND DISCUSSION

The glass-to-Kovar seal was examined in detail using the finite element analysis. The expansion differential and the setting temperature were assumed to be 10^{-4} in/in and 1000°F respectively. The following elastic properties were assumed:

$$E_g = 10^7 \text{ psi}, \qquad \nu_g = 0.22$$
$$E_m = 2 \times 10^7 \text{ psi}, \qquad \nu_m = 0.30$$

The glass/metal diameter ratio was taken as 2.0. The variation of axial stress in glass along the length of the seal is shown in Figs. 3 and 4 for five different aspect ratios. Figure 3 shows the stress at the interface (r = a) whereas Fig. 4 shows the stress at the outer surface (r = b).

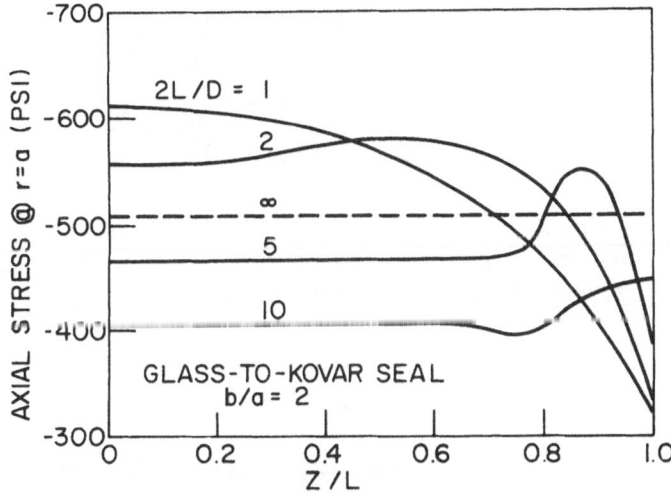

Fig. 3   Variation of Axial Stress in Glass
at r = a Along Seal Length: Glass-to-
Kovar Seal with b/a = 2

It should be noted that the axial stress varies not only along the length but also through the glass thickness. The latter variation is significant in the case of low aspect ratios. For example, for 2L/D = 1 the axial stress at mid-length is slightly over −600 psi at r = a and slightly below -100 psi at r = b. This huge variation must be taken into account by appropriate weighting technique in relating the tangential birefrin-

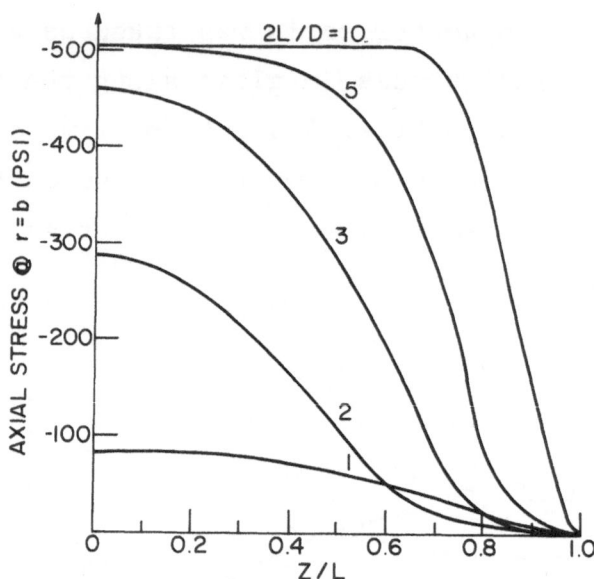

Fig. 4   Variation of Axial Stress in Glass
         at r = b Along Seal Length: Glass-to-
         Kovar Seal with b/a = 2

gence to axial stress and the expansion differential.
The tangential shape factor, eqn. 13, derived from Figs.
3 and 4 is shown in Fig. 5 for different values of moduli
ratio and glass/metal diameter ratio. The aspect ratio
of the seal was kept fixed at 10, the typical value of
a common laboratory seal. It is interesting to note
that the shape factor approaches 1 as the diameter ratio
goes to 1, i.e. as the glass layer gets thinner. This
is to be expected on physical grounds since the stress
variation through the glass thickness would be minimal
in the case of a thin glass layer. It should also be
noted that K_t decreases rather rapidly as the moduli

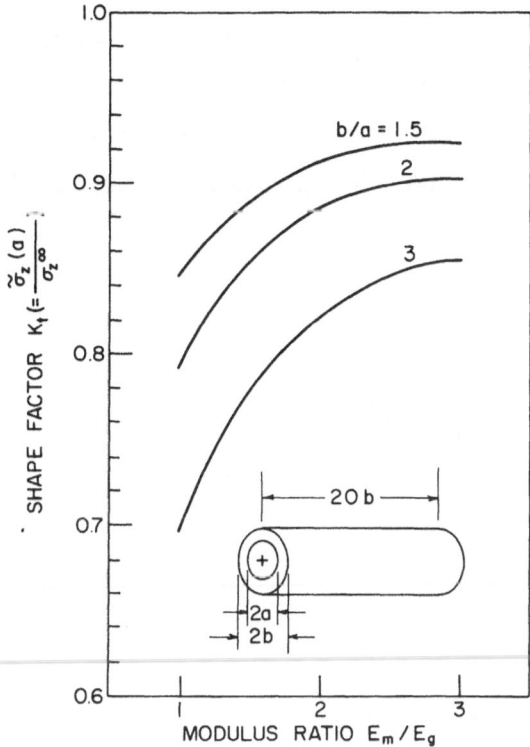

Fig. 5   Variation of Tangential Shape Factor with
         Moduli Ratio and Glass/Metal Diameter Ratio for
         Aspect Ratio of 10

ratio decreases and the diameter ratio increases. Thus
for seals made of similar materials with large diameter
ratio the error in computing expansion differential can
be large if the shape factor is not taken into account.

The axial shape factor is obtained by examining the
variation of circumferential and radial stresses at the
interface along the seal length. In an infinitely long
seal, these stresses do not vary along the seal length.

The shape factor, as defined by eqn. 17, was calculated
from the results of the finite element analysis of glass-
to-Kovar seal. The dependence of K_a on the aspect ratio
is shown in Fig. 6 for three different diameter ratios.
It should be noted that K_a deviates the most from unity

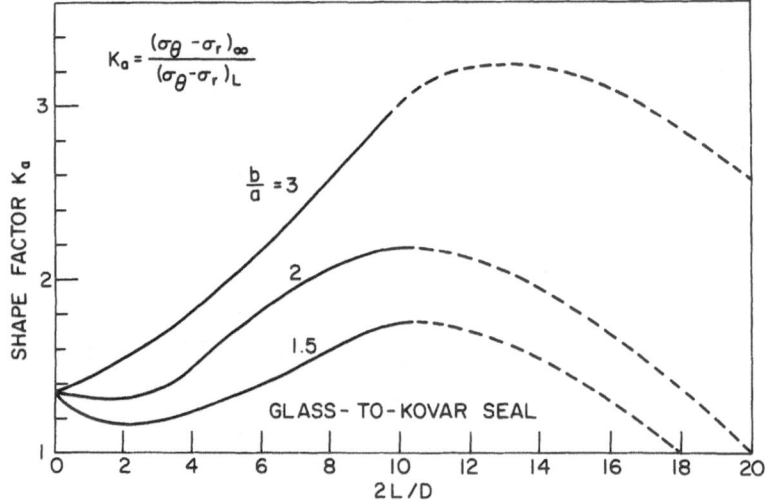

.Fig. 6 Variation of Axial Shape Factor with
        Aspect Ratio: Glass-to-Kovar Seal

for large values of b/a, i.e. when the glass thickness
is large, and the error in computing expansion differential can become large if K_a is not taken into account.
The finite element analysis was carried out for aspect
ratios of upto 10. The results are extrapolated thereafter and shown by the dotted line. It is clear that,
for very large aspect ratios, the shape factor should
approach unity.

EXPERIMENTAL VERIFICATION

An experimental glass-to-glass seal was made in the laboratory by inserting a 0.23 in. diameter cane of Corning Code 1720 glass in Code 9741 tubing with an I.D. of 0.24 in. and an O.D. of 0.33 inch. The physical properties of the two glasses are shown in Table I. The

Table 1. Physical Properties of Seal Members

(Nominal Values)

Property	Corning Code 1720	Corning Code 9741
E (10^6 psi)	12.7	7.2
ν	0.24	0.23
Set Point (°C)	672	413
Annealing Point (°C)	712	450
Softening Point (°C)	915	705
Expansion Coef. at 413°C (10^{-7}/°C)	45.4	50

cane-tubing assembly was heated to 750°C, above the softening point of the cladding glass, to form the seal. It was held at that temperature for ½ hr. (for annealing) and thereafter cooled at the rate of 1°C/min. to below

the strain point (400°C). It was then cooled to room
temperature at a faster rate.

The initial aspect ratio of the seal was 18. The
axial stress in the cladding was measured by tangential
viewing in an index-matching oil under polarized light.
The seal was sliced at both ends successively and the
birefringence was recorded, always at the mid-length
and at the interface. The observed birefringence was
normalized at an aspect ratio of 13.6, using a K_t value
of 0.9 predicted by the finite element analysis. The
data is shown in Fig. 7 along with finite element predictions for seals with b/a = 1.5 and two different
moduli ratios. The experimental seal had a b/a ratio

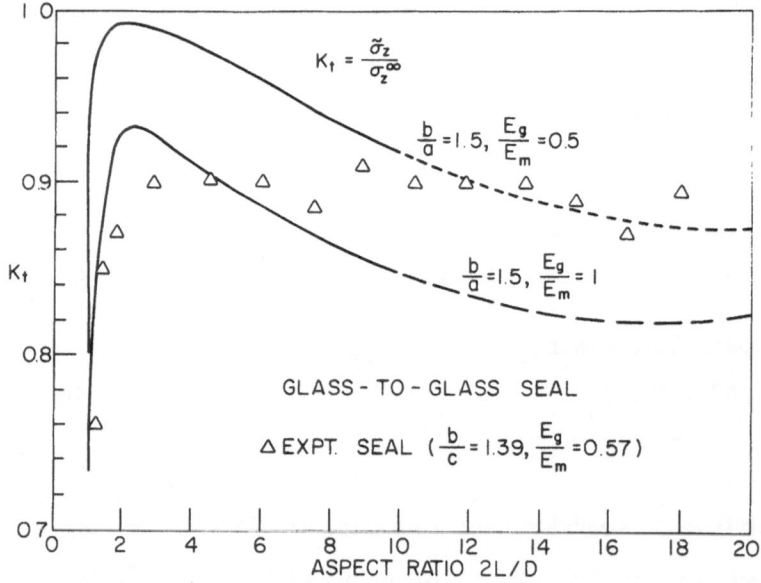

Fig. 7 Variation of Tangential Shape Factor
with Aspect Ratio for Glass-to-Glass
Seal

of 1.39 and a moduli ratio of 0.57. The data confirm
the trend predicted by finite element analysis. The
sharp peak predicted by FEA at low aspect ratios may not
be real in view of the coarse mesh used for the analysis.
The data clearly demonstrate that large errors can result in determining the expansion differential from
seals with low aspect ratios. It is therefore necessary
to correct the measured expansion differential by dividing it by the appropriate value of K_t.

The experimental seal, unfortunately, did not permit birefringence measurements in the axial direction
due to the presence of cord. It was also difficult to
align the light path along the interface. In order to
measure the principal stress difference $(\sigma_\theta - \sigma_r)$ it is
extremely important to have optical quality glass and a
well-aligned core-cladding boundary along the axial line
of sight. Efforts to produce such seals are continuing
at this laboratory.

CONCLUSIONS

In the case of short cylindrical bead seals the
variation of stresses through the glass thickness and
along the seal length requires a correction factor to
obtain the expansion differential accurately. In this
paper, two such factors - one for tangential viewing and
the other for axial viewing - are derived from the results of finite element analysis. The variation of the
shape factors, K_t and K_a, with aspect ratio and diameter

ratio is shown and the deviation of these factors from
unity is found to be greatest for large values of diameter ratio b/a and moduli ratio E_m/E_g. Thus the correction is most serious for seals of dissimilar materials, such as glass-to-steel, with relatively thick
glass layer. The predicted values of tangential shape
factor were verified by making a glass-to-glass bead
seal and examining the changes in birefringence as the
seal length was reduced. The agreement between theory
and experiment was found to be good. Based on this it
is felt that the shape factors predicted for glass-to-
Kovar and glass-to-steel seals are also reliable. The
discrepancy between theory and experiment at extremely
low values of aspect ratio is attributed to the coarse
mesh size used in the finite element analysis.

REFERENCES

1. A. W. Hull and E. E. Burger, "Glass-to-Metal Seals",
    Physics, Vol. 5, No. 12 (1934).

2. G. D. Redston and J. E. Stanworth, "Glass-to-Metal
    Seals", J. Soc. Glass Tech.; Vol. 29 (1945).

3. H. Rawson, "The Theory of Stresses in Two-component
    Glass to Metal Tube Seals", J. Sci. Instr.; Vol. 26,
    No. 1 (1949).

4. Hillel Poritsky, "Analysis of Thermal Stresses in
    Sealed Cylinders and the Effect of Viscous Flow
    During Anneal", Physics, Vol. 5, No. 12 (1934).

5. ASTM Standard F14   ; 1976 Annual Book of ASTM
    Standards, Part 43, Philadelphia, Pa.

6. R. C. O'Rourke, "Theoretical Studies in Photoelasti-
    city", Doctor's Thesis, Univ. of Michigan (1950).

REFERENCES (CONTINUED)

7. H. N. Ritland "Stress Measurements in Cylindrical Vessels", J. Am. Ceram. Soc.; Vol. 40, No. 5 (1957).

8. P. M. Sutton, "Stress Measurements in Circular Cylinders", J. Am. Ceram. Soc.; Vol. 41, No. 3 (1958).

9. Lamé, Leçons sur la theorie mathematique de l'elasticité des corps solides, Paris (1852).

10. M. Takagi*, "Theory of Dumet-to-Glass Sealing", Toshiba Review, TOREA, Vol. 5 (1950).

11. F. W. Martin, "Stresses in Glass-to-Metal Seals: I, The Cylindrical Seal", J. Am. Ceram. Soc.; Vol. 33, No. 7 (1950).

12. T. Kishii*, "Dumet-to-Glass Sealing", Toshiba Review, TOREA, Vo. 13 (1958).

13. T. Kishii*, "Stresses in Dumet-to-Glass Seals", Yogyo Kyokaio Shi, YGKSA, Vol. 66 (1958).

14. Gulati, S. T. and Hagy, H. E., "Theory of the Narrow Sandwich Seal", J. Am. Ceram. Soc. (to appear).

15. Gulati, S. T. and Hagy, H. E., "Finite Element Analysis of Narrow Sandwich Seal and Experimental Determination of Its Shape Factor", J. Am. Ceram. Soc. (to appear).

---

*   English translation available from SLA Translation Center, 35 W. 33rd St., Chicago, Ill. 60616, USA.

TRANSIENT (SUBSECOND) INTERFEROMETRIC TECHNIQUE FOR THERMAL

EXPANSION MEASUREMENTS AT HIGH TEMPERATURES*

A. P. Miiller** and A. Cezairliyan

National Bureau of Standards

Washington, DC   20234

ABSTRACT

The feasibility of a new interferometric technique for measuring thermal expansion of electrical conductors between room
temperature and temperatures in the range 1500 K and their melting
points is considered. The basic method involves rapidly heating
the specimen from room temperature to temperatures above 1500 K in
less than one second by the passage of an electrical current pulse
through it while simultaneously measuring the specimen temperature
and the shift in fringe pattern produced by a two-beam polarizing
interferometer; the specimen of rectangular cross section has two
opposite faces highly polished and is located as a double reflector
in the path of one beam. Preliminary measurements on a tantalum
specimen in the temperature range 1400 to 2300 K indicate that the
laser-based (He-Ne) interferometer in its initial configuration is
capable of measuring linear thermal expansion of a 3 mm specimen
with a sensitivity of better than 50 µm/m. The maximum error in
the reported values of thermal expansion is estimated to be not
more than 3%.

---

*This work was supported in part by the U.S. Air Force Office of
Scientific Research.

**Guest scientist from Brandon University, Brandon, Manitoba, Canada.

INTRODUCTION

Accurate measurements of thermal expansion based on frac-
tional-wave interferometry are usually restricted to temperatures
below about 1100 K. At higher temperatures, expansion measurements
are performed by means of push-rod dilatometry, x-ray diffractometry or telemicroscope methods. In all of these techniques, the
specimen is held at elevated temperatures for long periods of time
(minutes-to-hours) giving rise to problems associated with increased
heat transfer, evaporation, chemical reactions, etc., which become
particularly critical at temperatures above 2000 K.

The rapid (subsecond duration) pulse heating technique developed
recently at the National Bureau of Standards [1,2] for the measurement of selected thermophysical properties minimizes the problems
associated with exposing the specimen to high temperatures for
long periods. The technique has been successfully used to simultaneously measure heat capacity, electrical resistivity and thermal
radiation properties of a number of electrically-conducting refractory materials.

In this paper, the feasibility of adapting an interferometer
to the NBS pulse heating system is considered for the purpose of
measuring thermal expansion of metals between room temperature and
temperatures above 1500 K (up to their melting points).

The basic technique consists of pulse heating a specimen by
passing a subsecond-duration electrical current pulse through it
and simultaneously measuring the specimen temperature and the shift
in fringes produced by a two-beam interferometer in which a specimen dimension forms part of the light path. The specimen temperature is measured by means of a high-speed photoelectric pyrometer
[3] and is recorded digitally approximately every 0.8 ms with a
full scale resolution of about 1 part in 8000. In addition, the
pyrometer signal, together with the analog signal corresponding to
the fringe shift is recorded on a dual-beam oscilloscope. Thermal
expansion is determined from the measurement of the fringe shift
corresponding to a particular temperature.

The description of instrumentation given in this paper deals
primarily with the interferometric aspects of the measurement
system. Details regarding the construction and operation of the
pulse heating system are given elsewhere [1,2].

Tantalum was selected as the specimen material for the preliminary measurements reported herein on the basis of its high
melting point and other favorable properties at high temperatures,
as well as the availability of data in the literature for comparison
purposes.

INTERFEROMETRIC SYSTEM

A. The Interferometer

The interferometer, illustrated schematically in figure 1, is basically a Michelson interferometer with a polarizing He-Ne laser as the light source. The specimen, in the form of a long rectangular rod with two opposite faces highly polished, is used as a double plane mirror element in one of the two light paths thereby making the interferometer sensitive to a specimen dimension.

The linearly polarized beam from the laser is separated into two components by a polarizing beamsplitter (PB1). The component that passes straight through PB1 emerges with the plane of its electric field vector parallel to the plane of incidence (p-polarized). The other component which emerges from PB1 at right angles to the incident beam is also linearly polarized but with its electric field vector orthogonal to the plane of incidence (s-polarized).

The p-polarized component passes through a quarter-wave plate (QP1) which has its principal axis oriented at 45 degrees to the planes of polarization and so it emerges as a circularly polarized beam. Upon reflection at normal incidence from the 'front' surface of the specimen the sense of circular polarization is reversed so that after a second pass through QP1 the beam is again linearly polarized but now with s-polarization. This prevents the beam from returning to the laser; instead, it is reflected by PB1 towards a corner cube (C1) which returns the beam with sufficient translation that after another reflection by PB1 the beam bypasses the speciment. By similar consideration of the second combination of polarizing beamsplitter (PB2), corner cube (C2) and quarter-wave plate (QP2), one can show that after reflection from the 'back' surface of the specimen the component beam ultimately emerges from the interferometer with its original polarization, that is, p-polarization. Because of the successive front surface/back surface reflections, the optical path length followed by this beam is not affected by translational movements of the specimen arising from any mechanical vibrations that may be generated during the rapid pulse heating.

The s-polarized component, on the other hand, serves as a reference beam. After successive reflections by a pentaprism, corner cube C3, plane mirror M1 and beamsplitter PB2, the s-component beam is aligned with the p-component beam (from the specimen) by translational adjustments of C3 and angular adjustments of M1.

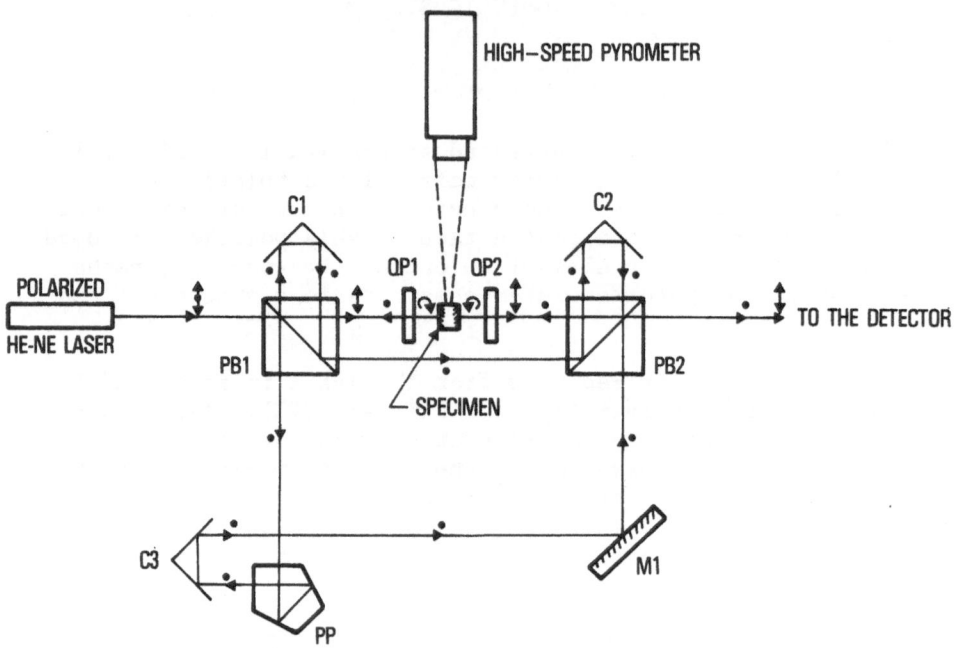

Figure 1. Schematic diagram of the interferometer consisting of
   the following optical elements: polarizing beamsplitters PB1,
   PB2; quarter-wave plates QP1, QP2; corner cubes C1, C2, C3;
   pentaprism PP; and plane mirror M1. The symbols ↕ and • refer
   to linear polarizations of the beam which are respectively parallel and perpendicular to the plane of the interferometer.
   Circular polarizations of the beam are indicated by the symbols
   ↻ and ↺ .

   The light output of the interferometer, then, consists of two
superimposed beams polarized mutually at right angles but with a
path-difference or phase difference between them. The path-
difference (PD) is the same at all points in the field of view
insofar as the interferometer can be adjusted to produce 'fringes
of infinite breadth'; in practice, the wavefronts of the two beams
do have slightly different curvatures so that very broad circular
fringes are seen instead. Interference between the two beams can
be readily observed by using an analyser with its principal axes at
45° relative to the planes of polarization; only the components of
the two electric field vectors resolved in the 45° direction are
transmitted by the analyser. The light intensity is essentially
uniform across the field of view becoming maximum and minimum respectively when

$$PD = n\lambda \quad \text{and} \quad PD = (n + 1/2)\lambda \tag{1}$$

where λ is the wavelength and $n = 0, 1, 2,\ldots$. Maximum contrast
between the bright and dark fringes occurs when the electric field
vectors of the output beams are equal, a condition which is easily
achieved by rotating the plane of polarization of the input beam
(from the laser) so that the relative amplitudes of the p- and
s-components will compensate for the unequal light losses along
the paths of the two beams.

As the specimen is rapidly heated through a temperature in-
crement ΔT, its 'length' ℓ changes by an amount $\Delta\ell$ giving rise to
a change in the path-difference of

$$\Delta PD = 2\Delta\ell \quad . \tag{2}$$

Also from (1) it is clear that

$$\Delta PD = \Delta n \lambda \tag{3}$$

where Δn represents the number of fringes (bright or dark) that
shift through the field of view. Combining (2) and (3) one can express the fractional linear expansion of the specimen as

$$\Delta\ell/\ell = \Delta n\lambda/2\ell \quad . \tag{4}$$

The cumulative fringe shift Δn, therefore, determines $\Delta\ell/\ell$ at each
temperature relative to the length at a selected reference temperature, usually room temperature.

The linear thermal expansion of metals between room temperature and their melting points is usually of the order of 2%. For
the values $\ell \simeq 3mm$ and $\lambda = 632.8$ nm used in this study, the corresponding Δn is of the order of 200 fringes. Since the fringe
shift occurs in a time interval of less than one second, a rapid
photodetection system is required to accurately count Δn.

B.   The Phase Quadrature Detector

There are a number of arrangements possible for deriving
signals in phase quadrature from the light output of interferometers so that bi-directional counting of the interference fringe
movements can be carried out [4,5]. The method considered here is
identical in principle to one developed very recently by Hocken[+]
at NBS. The basic components, shown schematically in figure 2,

[+] The authors gratefully acknowledge the helpful discussions with
R. J. Hocken, particularly, concerning the design of a phase
quadrature detector.

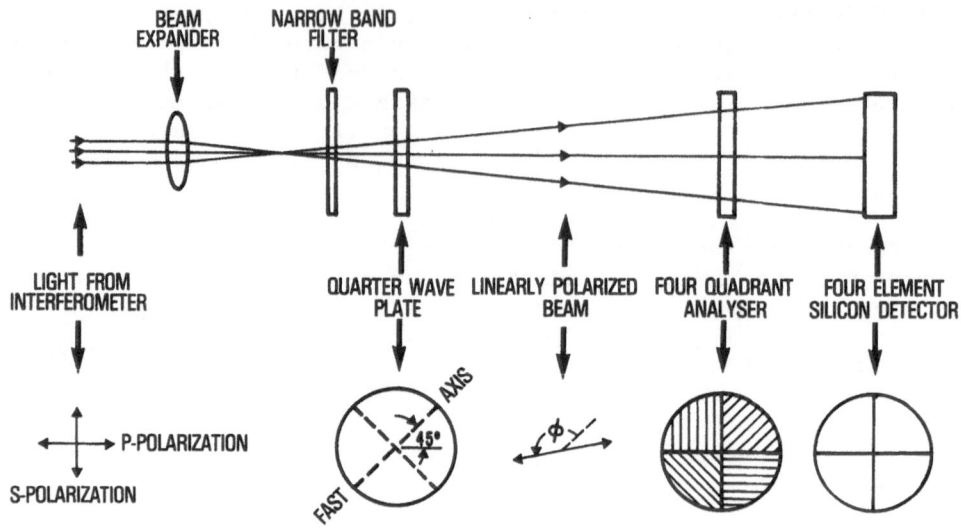

Figure 2.   Schematic diagram of the phase-quadrature detector.

consist of a beam expander, an interference filter, a quarter-wave
plate, a four-quadrant analyser and a detector consisting of four
silicon photodiodes, each converting the light passed by one
analyser quadrant into an electrical signal.

    The light radiated by the specimen during heating is largely
removed from the output beams of the interferometer by means of an
interference filter with a narrow passband (1 nm); the effect of
the residual background radiation passed by the filter is eliminated by differential amplification of the electrical signals from
the photodiodes[++].

    The quarter-wave plate with its principal ('fast') axis at 45°
to the planes of polarization converts each of the two linearly
polarized output beams from the interferometer into a circularly
polarized beam but with opposite senses of rotation. Since the
interferometer is adjusted so that the electric field amplitudes of
the output beams are equal, say E, the electric field vectors of
the two circularly polarized components have the same magnitude,
$E/\sqrt{2}$. The resultant or vector sum of the two (opposite) circular
polarizations is linearly polarized light with amplitude equal to
$2 \times E/\sqrt{2} = \sqrt{2}\,E$ and an angular orientation $\emptyset$ which is linearly
dependent on the phase difference δ between the two beams leaving

[++]The authors are indebted to M. S. Morse at NBS for the design and
construction of the electronics associated with the detector.

the interferometer. It can be shown that if $\emptyset$ is measured counter-clockwise from the direction of the 'fast' axis of the quarter-wave plate (see figure 2) then

$$\emptyset = \delta/2 \quad . \tag{5}$$

In other words, the polarization plane of the light reaching the analyser rotates through 180° whenever $\Delta\delta = 360°$; this corresponds to $\Delta PD = \lambda$ or $\Delta \ell = \lambda/2$ or, in terminology used earlier in the paper, a shift of one fringe.

The light intensity transmitted by a given analyser quadrant is proportional to the square of the electric field component in the direction of its principal axis. So if the directions of the principal axes in the first through fourth quadrants are at 0°, 45°, 90° and 135°, respectively, relative to the 'fast' axis direction (see figure 2), it follows that the transmitted light intensities will vary according to

$$I_1 \propto \cos^2\emptyset \quad , \tag{6}$$

$$I_2 \propto \cos^2(\emptyset - \pi/4) \quad , \tag{7}$$

$$I_3 \propto \sin^2\emptyset \quad , \tag{8}$$

$$\text{and} \quad I_4 \propto \sin^2(\emptyset - \pi/4) \quad . \tag{9}$$

In fact, the four electrical current signals generated by the four element detector will also have the same functional dependence on $\emptyset$ as expressed by equations (6) through (9), insofar as the detector response to light input is linear. Therefore, by differentially amplifying the current signals derived from diagonally opposite quadrants, one obtains two voltage signals which vary according to

$$e_A \propto \cos^2\emptyset - \sin^2\emptyset = \cos 2\emptyset \tag{10}$$

and

$$e_B \propto \cos^2(\emptyset - \pi/4) - \sin^2(\emptyset - \pi/4) = \sin 2\emptyset \tag{11}$$

Finally, these signals may be expressed in terms of the phase difference:

$$e_A = e_{max} \cos \delta \tag{12}$$

and

$$e_B = e_{max} \sin \delta \tag{13}$$

where e_{max} is the voltage amplitude. One cycle of either signal, then, corresponds to a shift of one fringe or specimen expansion of $\lambda/2$. The 90° phase difference between e_A and e_B makes them highly suitable for bi-directional counting, particularly in conjunction with an electronic up/down counter. For the purpose of this preliminary study, however, sufficient information about fringe movements is obtained by recording one of the signals on an oscilloscope trace photograph.

A more detailed description of the construction and operation of the interferometric system will be published elsewhere [6] when the capability of digitally recording the fringe shift count is added to the system.

MEASUREMENTS

A specimen of rectangular cross-section with nominal dimensions 3 x 3 x 76 mm long was fabricated from a cylindrical rod of tantalum (about 99.9% pure). Two opposite rectangular faces of the specimen were polished yielding a flatness of better than $\lambda/4$ at the center portions which reflect the light beam; the beam size of the He-Ne laser (2 mW) used for the interferometer was about 1 mm in diameter. A third surface, used for the pyrometric temperature measurements, was also polished.

The pyrometer response was optimized by dividing the temperature interval of the measurements (1400 to 2300 K) into four ranges. Two experiments were performed in each temperature range. Prior to each experiment, we adjusted a resistance in series with the specimen in order to limit the maximum temperature achieved by the specimen during the pulse heating; the specimen was then heated in a vacuum environment of 1.3 mPa ($\sim 10^{-5}$ torr) from room temperature to the desired temperature by means of an electrical current pulse of 850 ms duration. Upon completion of the experiments, we calibrated the high-speed pyrometer [3] using a tungsten filament reference lamp which, in turn, had been calibrated by the Temperature Measurement Section of the National Bureau of Standards.

Figure 3 is a photograph of the two traces of a dual-beam oscilloscope. The upper trace shows the time variation of the radiance of the specimen surface as seen by the pyrometer, as the specimen is heated from room temperature to the maximum temperature. The dots forming three long horizontal lines in the pyrometer output correspond to the radiance from a reference source when viewed directly (the uppermost horizontal line) and through two neutral filters of different densities. The lower trace shows the interferometer fringe shift produced by the heating of the specimen. Each cycle of the interferometer output indicates a shift of one

fringe in the interference pattern and a specimen expansion of one-
half wavelength.

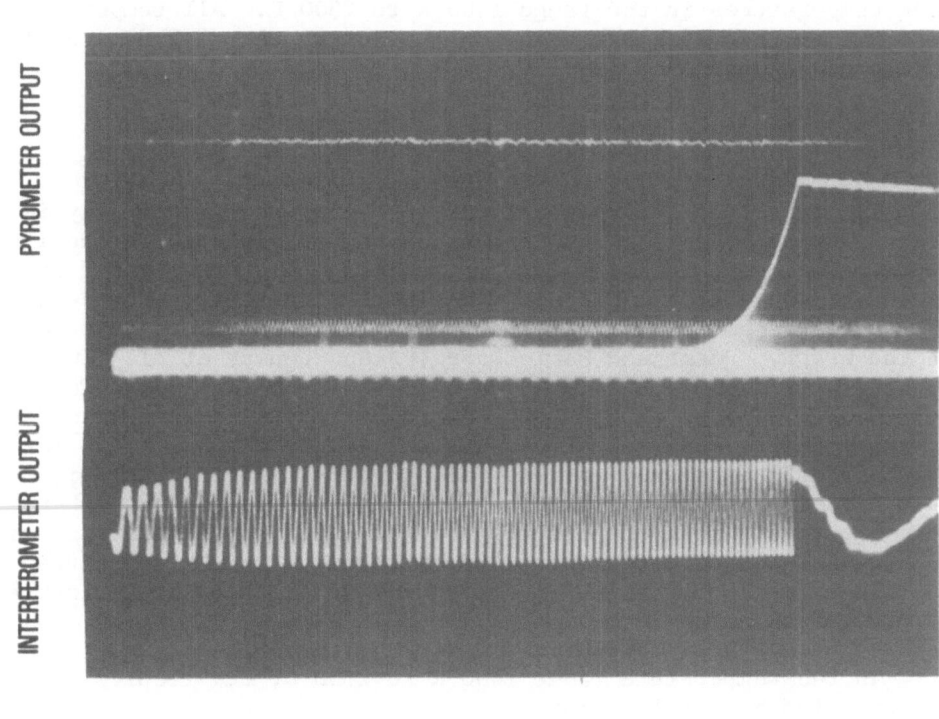

TIME

200ms

Figure 3. Oscilloscope trace photograph of the specimen radiance
(as observed by the pyrometer) and the specimen expansion (as measured by the interferometer) during a typical pulse heating experiment. Dots forming the horizontal lines in the pyrometer
output correspond to the radiance from a reference source. Each
cycle of interferometer output corresponds to an expansion of
$\lambda/2$.

For each experiment in a given temperature range the cumula-
tive fringe shift, hence the cumulative thermal expansion $\Delta\ell/\ell$ was
determined at those instants of time on the oscillogram when the
specimen surface radiance became equal to 'radiances' from the

reference source; the corresponding radiance temperatures[#] of the specimen were determined by the pyrometer calibration. Using the normal spectral emittance data of Malter and Langmuir [7], we converted the specimen radiance temperatures into true (blackbody) temperatures; the temperature conversion involved adding 65 degrees up to 190 degrees to the measured radiance temperatures to yield true temperatures in the range 1400 K to 2300 K. All temperatures reported in this work are based on the International Practical Temperature Scale of 1968 [8].

RESULTS AND DISCUSSION

The results of the present work are compared graphically in figure 4 with linear thermal expansion values reported in the literature. Each of our data points represents the average value of $\Delta\ell/\ell$ determined from a pair of experiments; the difference between values from a given pair of experiments is too small to be resolved on the graph. Rasor and McClelland [9] and Conway et al. [10] obtained their expansion data using telemicroscope methods. The values of expansion reported by Edwards et al. [11] were measured by x-ray diffraction techniques whereas Amonenko et al. [12] obtained their data using conventional dilatometry. Our results lie about midway between the extremes of data reported by the other investigators throughout the temperature range of our experiments and are in good agreement with the solid curve which represents the linear expansion values recommended for tantalum by Touloukian, et al. [13].

In the temperature range 1400 K to 2300 K, our results can be approximated by the function

$$\Delta\ell/\ell = -0.1410 + 5.613 \times 10^{-4}T + 7.434 \times 10^{-8}T^2 \tag{14}$$

where T is the absolute temperature and the numerical coefficients are those obtained from a least-squares fit to our data (average standard deviation = 0.2%; 16 data points). Figure 5 presents the deviation of results yielded by individual experiments from the smooth function represented by equation (14). The dashed lines represent the uncertainty in $\Delta\ell/\ell$ that would arise from an uncertainty in fringe count of 0.5 fringe (an uncertainty in specimen expansion of $\lambda/4$); this would correspond to a measurement sensitivity of 50 μm/m for our 3 mm specimen. The deviation plot indicates that the imprecision of our measurements is smaller than the un-

---

[#]

Radiance temperature (or brightness temperature) is the apparent temperature of the specimen surface corresponding to the effective wavelength of the measuring pyrometer.

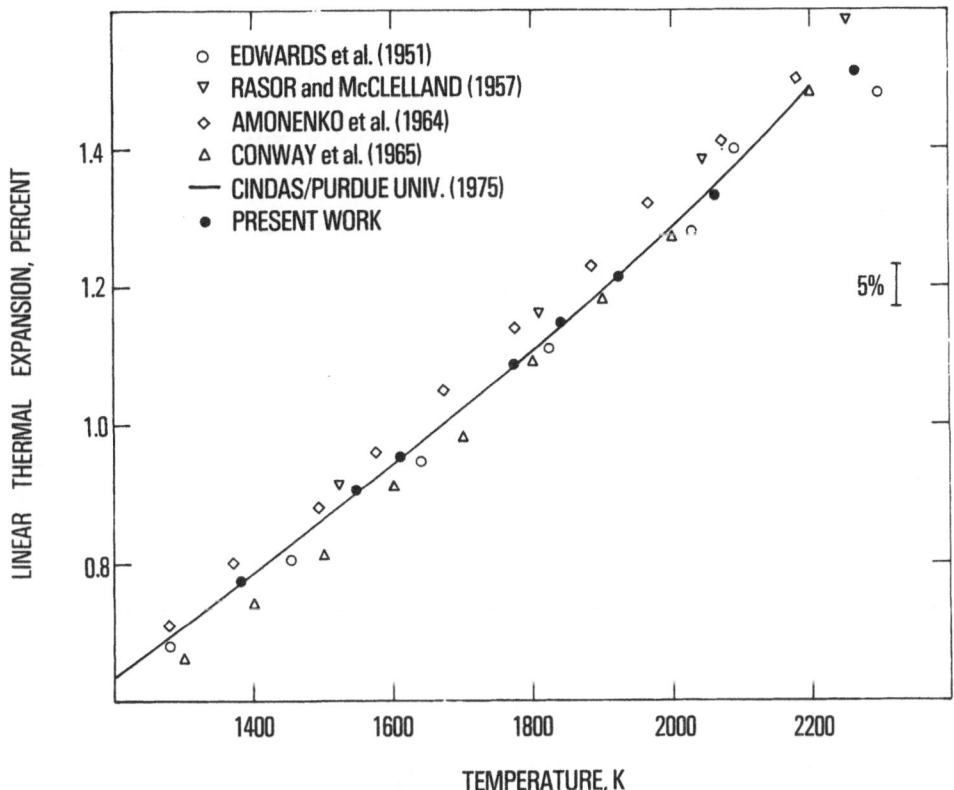

Figure 4. Linear thermal expansion of tantalum reported in the literature.

certainty represented by the dashed lines by about a factor of 2; hence the sensitivity of our measurements in this feasibility study is better than 50 μm/m.

A major source of error in our results arises from the conversion of the measured radiance temperatures to true temperatures. The uncertainty in the normal spectral emittance is believed to be not greater than 10% which, at 2000 K, corresponds to an error in true temperature of about 20 degrees or about 1%. The fringe shift count is believed to be accurate to within 0.5 fringe which, at 2000 K, corresponds to an error in $\Delta \ell / \ell$ of about 0.5%. An additional uncertainty of about 1% comes from the measurement of the specimen dimensions at room temperature. Therefore, the maximum error in our reported values of linear expansion is estimated to be not more than 3%.

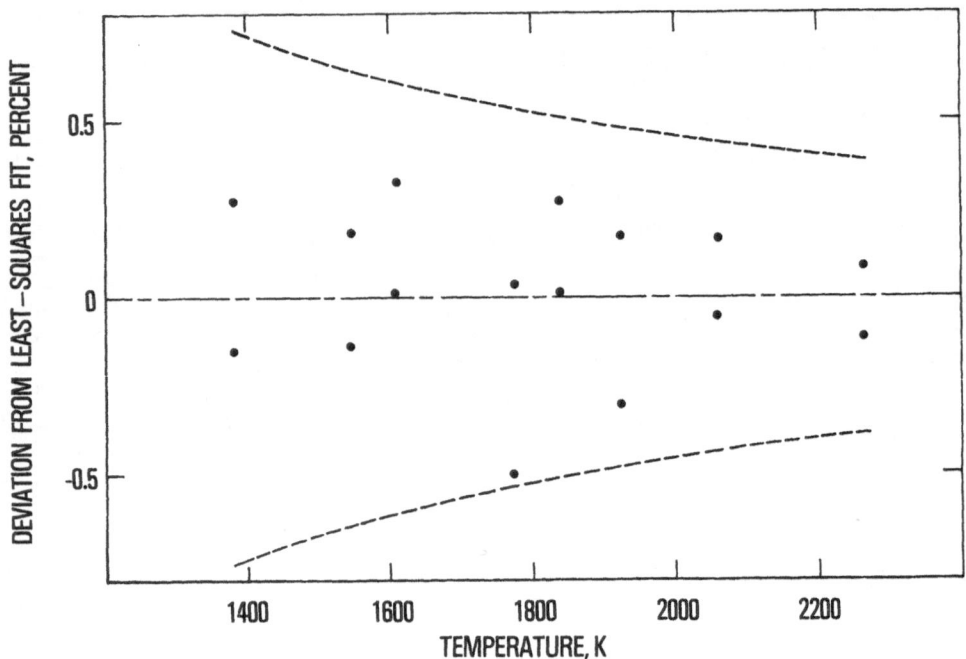

Figure 5. Deviation of results yielded by individual experiments
from the smooth function represented by equation (14). The
dashed lines represent the uncertainty in $\Delta \ell / \ell$ that would arise
from an uncertainty in specimen expansion of $\lambda / 4$.

In conclusion, the results obtained in this study demonstrate
the feasibility of the transient interferometric technique for
measuring thermal expansion of metals at high temperatures. The
addition of a digital up/down counter to the measurement system
should permit accurate and rapid counting of the fringe shifts up
to the melting points of specimens. The accuracy of the technique
can be further improved by using specimens containing a blackbody
cavity so that the true temperature can be measured directly.

REFERENCES

1. A. Cezairliyan, M. S. Morse, H. A. Berman and C. W. Beckett,
    "High-Speed (Subsecond) Measurement of Heat Capacity, Electri-
    cal Resistivity, and Thermal Radiation Properties of Molybdenum
    in the Range 1900 to 2800 K," J. Res. Nat. Bur. Stand. (U.S.)
    74A, (Phys. and Chem.) 65-92 (1970).

2. A. Cezairliyan, "Design and Operational Characteristics of a
    High-Speed (Millisecond) System for the Measurement of

Thermophysical Properties at High Temperatures," J. Res. Nat. Bur. Stand. (U.S.) 75C, (Eng. and Instr.) 7-18 (1971).

3. G. M. Foley, "High-Speed Optical Pyrometer," Rev. Sci. Instr. 41, 827-834 (1970).

4. W. R. C. Rowley, "Some Aspects of Fringe Counting in Laser Interferometers," IEEE Transactions on Instrumentation and Measurement, IM-15, 146-149 (1966).

5. J. Dyson, in, "Interferometry as a Measuring Tool," Hunt Barnard Printing Ltd., Aylesbury, 206 pp. (1970).

6. A. P. Miiller and A. Cezairliyan, "Design and Operational Characteristics of an Interferometric System for the Transient (Subsecond) Measurement of Thermal Expansion at High Temperatures," in preparation.

7. L. Malter and D. B. Langmuir, "Resistance, Emissivities and Melting Point of Tantalum," Phys. Rev. 55, 743-747 (1939).

8. International Practical Temperature Scale of 1968, Metrologia 5, 35-44 (1969).

9. N. S. Rasor and J. D. McClelland, "Thermal Properties of Materials. Part I. Properties of Graphite, Molybdenum, and Tantalum to their Destruction Temperatures," U.S. Air Force Rept. WADC-TR-56-400, Pt. I, 53 pp. (1957).

10. J. B. Conway, R. M. Fincel, Jr., and A. C. Losekamp, "Effects of Contaminants on the Thermal Expansion of Tantalum," Trans. Met. Soc. AIME 233 (4), 841-842 (1965).

11. J. W. Edwards, R. Speiser and H. L. Johnston, "High Temperature Structure and Thermal Expansion of Some Metals as Determined by X-Ray Diffraction Data. I. Platinum, Tantalum, Niobium, and Molybdenum," J. Appl. Phys. 22 (4), 424-428 (1951).

12. V. M. Amonenko, P. N. Vyugov and V. S. Gumenyuk, "Thermal Expansion of Tungsten, Molybdenum, Tantalum, Niobium, and Zirconium at High Temperatures," High Temp. (USSR) 2 (2), 22-24 (1964).

13. Y. S. Touloukian, R. K. Kirby, R. E. Taylor and P. D. Desai, "Thermophysical Properties of Matter," Vol. 12, Thermal Expansion, IFI/Plenum, New York, 1975.

UNIVERSAL GRAPHITE DILATOMETER

FOR HIGH TEMPERATURE STUDIES

Peter S. Gaal

Anter Laboratories, Inc.

1700 Universal Road Pittsburgh, PA 15235 USA

ABSTRACT

A new, universal expansion measuring device has been developed for the accurate and rapid measurements of thermal expansion to very high temperatures. The design and operations are described in detail and results are presented of measurements on a number of high temperature materials.

The instrument is unique in its capability of allowing the investigator to employ a recording graphite dilatometer concurrently with an optical extensometer trained directly on the specimen. With this combination, an accuracy not heretofore possible for dilatometers is attained without sacrificing the advantages offered by continuous recording. The device may be calibrated by absolute measurements, independent of any standards or reference materials. The unit is equipped with its own furnace, power supply and fully digital linear programmer as well as analog and digital data acquisition systems.

1.0 INTRODUCTION

Advances in technology during the past decade have brought on an unprecedented upsurge in materials development and with it, the everpresent requirements for their characterization. With the extension of the operating range well into the above 1500°C region, many of the classical measuring methods found themselves without usable reference standards. Thermal expansion testing was among these that had to look toward an absolute scheme to replace or to complement the conventionally used comparative process.

THE OPTICAL EXTENSIOMETER HAS BEEN IN USE NEARLY A CENTURY, AND IT IS AMONG THE MORE OR LESS ABSOLUTE METHODS. BUT IT HAS BEEN POINTED OUT BY SEVERAL INVESTIGATORS THAT EVEN THAT CAN BE SUBJECT TO SEVERE LIMITATIONS AND ERRORS IF NOT PERFORMED WITH THE UTMOST CARE. DILATOMETRY, THE OLD STANDBY IS EQUALLY VULNERABLE, HOWEVER, SINCE IT IS A COMPARATIVE METHOD WHICH DEPENDS ON KNOWING THE CHARACTERISTICS OF A REFERENCE STANDARD. THERE ARE NO STANDARDS AND EVEN REFERENCE MATERIALS THAT WORK IN A GRAPHITE ATMOSPHERE ARE AVAILABLE ONLY THROUGH ROUND-ROBIN PROGRAMS.

THE USE OF A REFERENCE STANDARD TO CALIBRATE A DILATOMETER PRESUPPOSES THAT THE DILATOMETER IS THERMALLY STABLE AND WILL NOT CHANGE ITS OWN EXPANSION CHARACTERISTICS WITH TIME, ETC. ALTHOUGH THIS IS A REASONABLE PREMISE, SOME DOUBT STILL EXISTS ESPECIALLY IF SPECIMEN LENGTHS OTHER THAN THE STANDARD, OR HEATING RATES DIFFERING GREATLY FROM WHAT WAS USED DURING THE CALIBRATION ARE EMPLOYED. NEVERTHELESS, EVEN THESE QUESTIONS DO NOT MINIMIZE THE UTILITY OF A DILATOMETER SYSTEM IN PROVIDING A CONTINUOUS CURVE AND USUALLY IN AN AUTOMATIC MODE.

THE HIGH TEMPERATURE DILATOMETER SYSTEM DESCRIBED IN THIS PAPER IS AIMED AT COMBINING THE AFOREMENTIONED ADVANTAGES OF BOTH METHODS. IT OFFERS THE INVESTIGATOR A CONTINUOUSLY RECORDING DEVICE OPERATING UP TO 3000°C THAT CAN BE OPTICALLY CROSS-CHECKED AT ANY POINT IN THE RUN, THUS ELIMINATING THE RELIANCE ON SECONDARY STANDARDS AND ERASING ALL DOUBTS ABOUT THE ABSOLUTE ACCURACY OF THE DATA.

2.0   DESCRIPTION

2.1   FURNACE

IN ORDER TO PROVIDE A UNIFORM SPECIMEN TEMPERATURE, A SPECIAL FURNACE WAS DESIGNED (FIG. I). A GREAT DEAL OF ATTENTION WAS GIVEN TO OTHER RELATED PARAMETERS, SUCH AS LOW THERMAL MASS AND THIN WALLS. THE FIRST ONE WAS NEEDED TO ALLOW RAPID HEAT-UPS AND COOL-DOWNS WHILE THE SECOND WAS NECESSITATED BY THE COMPARATIVELY SHORT-WORKING DISTANCE TELEMICROSCOPES. THE DESIGN UTILIZES TO MAXIMUM BENEFIT THE HIGHLY ANISOTROPIC NATURE OF PYROLITIC GRAPHITE. (HIGHLY CONDUCTIVE IN THE WITH GRAIN DIRECTION AND IS A VERY GOOD INSULATOR IN THE CROSS GRAIN DIRECTION.) A TUBULAR CONFIGURATION SURROUNDING THE HEATING ELEMENT WAS CHOSEN, WHICH TENDS TO EVEN OUT THE HOT ZONE AXIALLY WHILE ACTING AS A VERY EFFECTIVE RADIAL THERMAL BARRIER RADIALLY. THE SUPPORTING STRUCTURES FOR THE PYROGRAPHITE ARE FABRICATED FROM COARSE GRAIN GRAPHITE AND ARE FURTHER INSULATED WITH CARBON FELT AND IN PLACES LOW DENSITY CARBON FOAM.

THE OUTER SHELL OF THE FURNACE IS MACHINED OUT OF A STAINLESS STEEL PIPE WITH A WELDED INNER SHELL. DEVIATING FROM THE CONVEN-

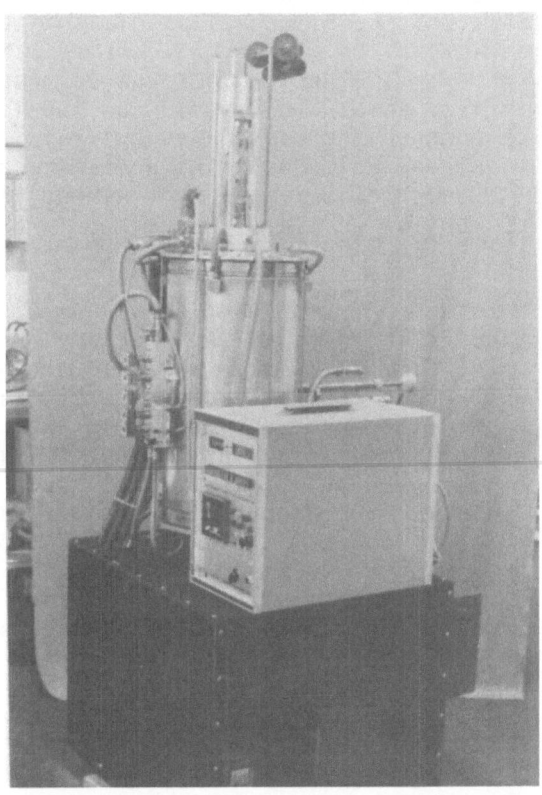

FIGURE I

ULTRA-HIGH TEMPERATURE DILATOMETER

TIONAL, DOUBLE WALL CONSTRUCTION, SOLID SIGHT PORT BLOCKS WERE
CHOSEN WITH FIVE VIEW PORTS ON EACH FACE OF THE FURNACE. ACTUAL
TESTS SHOWED THAT NO ADVERSE HEAT BUILD-UP IS PRESENT IN THE PORT
AREAS. THIS PATTERN ALSO AIDED IN PROVIDING A SHORT OPTICAL PATH
BETWEEN THE SPECIMEN AND THE TELESCOPE, AND ALLOWED A READY DE-
TERMINATION OF THE THERMAL GRADIENT ALONG THE SAMPLE.

THE TOTAL DISTANCE BETWEEN THE FRONT FACE OF THE SIGHT GLASS AND
THE AXIS OF THE FURNACE IS 4.75 INCHES, PERMITTING THE USE OF
TELESCOPES WITH AS HIGH AS 20X MAGNIFICATION EVEN WITHOUT A RELAY
OPTICS. THE TUBULAR GRAPHITE HEATING ELEMENT IS INSERTED AXIALLY
WITH FINGER CONTACT TYPE SLIDING CONNECTION ON THE TOP AND THE
BOTTOM. THE INSIDE BORE OF THE ELEMENT IS 2 1/8 IN., PROVIDING
NEARLY 9 INCHES OF UNIFORM HOT ZONE. THE HOLE PATTERN AND THE
WALL THICKNESS IS ADJUSTED EXPERIMENTALLY (TRIMMED) TO ACCOUNT
FOR UNEVEN DISTRIBUTION IN THE VERTICAL SYSTEM.

2.2 DILATOMETER

THE DILATOMETER (FIG. 2) IS INSERTED FROM THE TOP AND IT PROTRUDES
INTO THE HEATED SPACE (FIG. 3) IN SUCH A FASHION SO AS TO HAVE
THE TOP AND THE BOTTOM EDGE OF THE SPECIMEN VISIBLE THROUGH THE
SIGHT PORTS. A UNIQUE FEATURE OF THE DEVICE IS THE PROVISION FOR
LOADING THE DILATOMETER ITSELF FROM THE BOTTOM. (THIS PORTION
OF THE DILATOMETER IS FORMED BY A THREADED PLUG THAT IS SCREWED
INTO THE TUBE.) THIS WAY, THE UPPER INSTRUMENT SECTION IS NOT
DISTURBED DURING SAMPLE CHANGES. THE EXPANSION IS TRANSMITTED VIA
A CENTRAL PUSHROD THAT IS GROUND AND POLISHED AND IS GUIDED ONLY
AT TWO POINTS TO MINIMIZE FRICTION. THE DILATOMETER TUBE IS
FIXED TO A WATER-COOLED BASE CONSTRUCTED OF HEAT-TREATED THERMAL-
LY STABILIZED INVAR. THE TRANSDUCER SECTION IS MOUNTED ON TOP OF
THIS SECTION. A QUARTZ EXTENSION ROD IS FASTENED TO THE GRAPHITE
PUSHROD TO MINIMIZE HEAT CONDUCTION TO THE TRANSDUCER CORE WHILE
NOT ADDING APPRECIABLY TO THE SYSTEM EXPANSION. THE JOINT BETWEEN
THE GRAPHITE PUSHROD AND THE QUARTZ IS WATER-COOLED. CAREFULLY
CONTROLLED EXPERIMENTS WITH TEMPERATURE-SENSITIVE INDICATING
PAINTS HAVE PROVED THAT THE GAUGE SECTION ABOVE THE QUARTZ REMAINS
AT THE SAME TEMPERATURE AS THE REST OF THE WATER-COOLED HEAD.
THE PUSHROD IS STATICALLY COUNTERBALANCED. THE DESIGN ALLOWS
RAPID LOADING AND UNLOADING AS WELL AS A HIGH FLEXIBILITY TO VARY
THE SAMPLE LENGTH WITHOUT THE NEED OF RECALIBRATION. THIS IS
ACCOMPLISHED BY A MOTORIZED TRANSLATION STAGE TO MOVE THE TRANS-
DUCER (INDICATED BY A DIAL GAGE). THE ENTIRE DILATOMETER IS
MACHINED OUT OF A SINGLE BLOCK OF LOW EXPANSION GRAPHITE CLOSELY
OBSERVING GRAIN ORIENTATION AND THE RELATIONSHIP OF VARIOUS PARTS
TO EACH OTHER.

THE MECHANICAL MOTION OF THE PUSHROD IS TRANSLATED INTO AN EQUIV-
ALENT ELECTRICAL SIGNAL BY A LINEAR VARIABLE DIFFERENTIAL TRANS-
FORMER (LVDT). THE SIGNAL FROM THE TRANSDUCER IS PASSED THROUGH

FIGURE 2

MOTORIZED DILATOMETER HEAD WITH COUNTERBALANCING

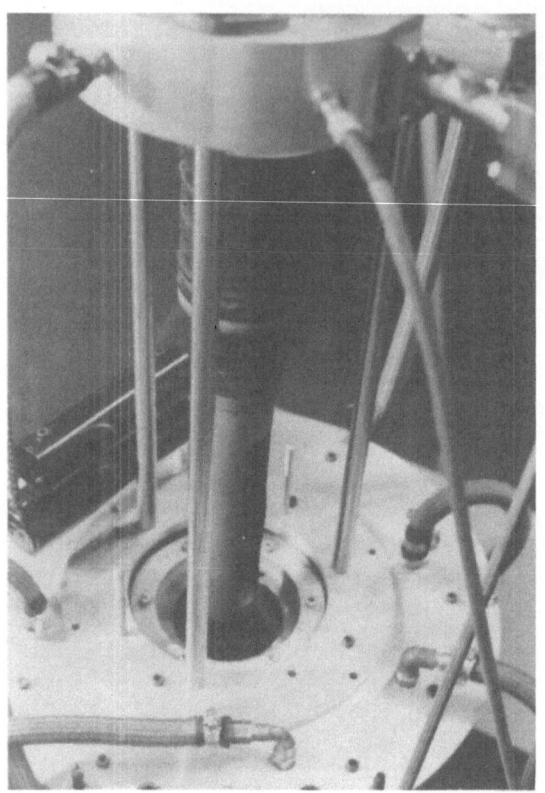

FIGURE 3

GRAPHITE DILATOMETER
(SHOWN WITH HEAD RAISED) WITH SPECIMEN CAVITY

CONDITIONING NETWORKS TO PROVIDE CALIBRATION AND ZERO SHIFT. IT
IS THEN DISPLAYED AND RECORDED DIGITALLY AND ALSO ON AN X-Y
RECORDER. THE DEVICE HAS PROVISIONS TO AUTOMATICALLY COMPENSATE
FOR THE LENGTH OF THE SAMPLE ALLOWING THE RECORDING OF PERCENT
EXPANSION DIRECTLY.

2.3 CENTRAL PROGRAMMER

THE TEMPERATURE OF THE FURNACE IS CONTROLLED BY AN SCR POWER SUP-
PLY IN CONJUNCTION WITH A DIGITAL TEMPERATURE PROGRAMMER, AND A
THREE-MODE CONTROLLER.

THE HEART OF THE SYSTEM IS THE FULLY DIGITAL UNITHERM^{TM} MODEL 150
CENTRAL PROGRAMMER THAT IS USED TO COORDINATE ALL FUNCTIONS WITH-
IN THE GROUP. BY DELEGATING ALL CONTROL RESPONSIBILITIES TO ONE
UNIT, A PERFECT SYNCHRONIZATION OF ALL OPERATIONS IS ACHIEVED IN
AN ECONOMICALLY ATTRACTIVE WAY. THE CIRCUITS ARE OF ADVANCED
DESIGN, EMPLOYING THE LATEST TTL INTEGRATED CIRCUIT TECHNIQUES
AND ARE BUILT ON PLUG-IN CIRCUIT BOARDS, ALLOWING EASY FIELD RE-
PAIRS OR CHANGES, IF NECESSARY.

THE PROGRESS OF THE CONTROL POINT (TEMPERATURE) IS DISPLAYED
DIGITALLY DIRECTLY IN °C. THIS IS MADE POSSIBLE BY THE USE OF
LINEARIZED THERMOCOUPLE AND PYROMETER CIRCUITS. THE DIGITAL
DISPLAY AND SEVERAL OTHER STATUS INDICATORS PROVIDE THE OPERATOR
WITH A COMPLETE PICTURE AT ALL TIMES.

OPERATING MODES ARE ESSENTIALLY ANY COMBINATION OF DELAYED START,
HEAT, SOAK, COOL, AND AUTOMATIC RECYCLE, AND A UNIQUE STEP-RAMP
THAT CAN YIELD HIGHLY ACCURATE EQUILIBRIUM DATA, AND IS ESPECIAL-
LY SUITED FOR OPTICAL CROSS MEASUREMENTS. THE FOLLOWING SPECIFIC
PROGRAMMING FORMS ARE READILY ATTAINED AND REPRODUCED WITH EX-
TREME ACCURACY:

A) LINEAR HEAT-UP, AND SHUT-OFF AT A PREDETERMINED TEMPERATURE.
B) LINEAR HEAT-UP TO A PREDETERMINED TEMPERATURE, AND LINEAR
    COOL-DOWN.
c) SAME AS A), EXCEPT WITH A PRESET SOAK PERIOD AT MAXIMUM TEM-
    PERATURE.
D) SAME AS B), EXCEPT WITH A PRESET SOAK AT MAXIMUM TEMPERATURE.
E) SAME AS A) OR B), EXCEPT WITH A PRESET DELAYED START-UP.
F) SAME AS ANY ONE OF THE ABOVE, EXCEPT AUTOMATICALLY RECYCLED.
G) STEPWISE HEATING AT MAXIMUM RATE IN 100° INTERVALS. A PRESET
    SOAK IS INSTITUTED AT EACH TEMPERATURE. AFTER THE SOAK, THE
    CYCLE PROCEEDS TO THE NEXT TEMPERATURE AUTOMATICALLY.

TYPE A), B), E), AND F) CYCLES ARE AVAILABLE IN THIS LAST OPERAT-
ING MODE.

IN ADDITION TO THE STANDARD CYCLE COMPONENTS, THERE ARE TWO

FORCED DWELLS INCORPORATED AT THE PEAK (10 MINUTES) AND AT THE END (3 HOURS) OF THE CYCLE TO ENSURE THAT THERMAL EQUILIBRIUM HAS BEEN ACHIEVED BEFORE CONTINUATION OR RECYCLE.

2.3.1 DIGITAL CLOCK

IN CASES WHERE KINETIC PROPERTIES ARE BEING DETERMINED (SINTERING OR PROCESS STUDIES), IT IS ESSENTIAL TO HAVE A GOOD TIME REFERENCE RECORDED RIGHT ALONG WITH THE DATA. FOR THIS PURPOSE, A DIGITAL CLOCK MODULE WAS ADDED TO PROVIDE ACCURATE TIME MARKERS.

2.3.2 DATA RECORD CONTROL

TO ENSURE COMPLETELY SYNCHRONOUS OPERATION AND TRUE CORRESPONDENCE BETWEEN VARIABLES, THE RECORDING CYCLES (REGARDLESS OF BEING ANALOG OR DIGITAL) ARE INITIATED AND GOVERNED BY SIGNALS DERIVED FROM THE CENTRAL PROGRAMMER. RECORDING CYCLES (SUCH AS PRINT OR PUNCH) ARE SET TO 10, 2, 1 CYCLE/MINUTE AND 2, 1 CYCLE/10 MINUTE INCREMENTS (SWITCH SELECTABLE).

3.0 THEORY OF OPERATION

LINEAR THERMAL EXPANSION IS THE CHANGE IN LENGTH PER UNIT LENGTH RESULTING FROM A CHANGE IN TEMPERATURE OF THE MATERIAL. THE MOST COMMONLY USED MEASURE OF THIS EXPANSION IS THE MEAN COEFFICIENT, $\bar{\alpha}$, WHICH IS DEFINED:

$$\bar{\alpha} = \frac{L - L_0}{L_0 (T - T_0)} = \frac{\Delta L}{L_0 \Delta T} \qquad (1)$$

FOR THE TEMPERATURE RANGE T_0 TO T. L_0 AND L ARE THE INITIAL AND FINAL SPECIMEN LENGTHS, RESPECTIVELY.

A DILATOMETER ALWAYS MEASURES THE DIFFERENCE BETWEEN THE THERMAL EXPANSION OF THE SPECIMEN AND THE MATERIAL OF THE DILATOMETER ITSELF.

$$(\Delta L/L)_{SAMPLE} = (\Delta L/L)_{MEASURED} + (\Delta L/L)_{DEVICE} \qquad (2)$$

WHERE ΔL IS THE TOTAL EXPANSION AT ANY GIVEN TEMPERATURE T IN ORDER TO DETERMINE THE MAGNITUDE OF THE $(\Delta L/L)_{DEVICE}$ TERM, A

WELL-KNOWN THERMAL EXPANSION (HENCEFORTH REFERRED TO AS STANDARD) MUST BE TESTED. THE SAME EQUATION WILL HOLD TRUE FOR THE STANDARD AS WELL, BUT FOR CLARITY, THE SUBSCRIPTS ARE CHANGED:

$$(\Delta L/L)_{STANDARD} = (\Delta L/L)_{DEVIATION} + (\Delta L/L)_{DEVICE} \qquad (3)$$

TABLE I

LINEAR THERMAL EXPANSION OF SELECTED GRAPHITES

TEMPERATURE	PERCENT EXPANSION		
°C	ATJ(WG)*	ATJ(AG)*	AXM5Q**
25	0.000	0.000	0.000
1200	.390	.420	.915
1400	.475	.540	1.090
1600	.560	.630	1.280
1800	.650	.740	1.472
2000	.745	.860	1.675
2200	.850	.980	1.885
2400	.960		2.110
AVERAGE NUMBER OF TESTS	6	10	4
AVERAGE DEVIA= TION AT 2000°C	±.02	±.02	±.015

* PRODUCT OF UNION CARBIDE CORP.
** PRODUCT OF POCO GRAPHITE CO.

THE TERM ON THE LEFT SIDE IS A KNOWN QUANTITY; THE "DEVIATION" TERM IS WHAT IS MEASURED, LEAVING ONLY THE "DEVICE" TERM UNKNOWN. BY COMBINING EQUATIONS 2) AND 3), THE WORKING EQUATION FOR THE ORDINARY DILATOMETER IS OBTAINED:

$$(\Delta L/L)_{SAMPLE} = (\Delta L/L)_{MEASURED} + (\Delta L/L)_{STANDARD} - (\Delta L/L)_{DEVIATION} \quad (4)$$

IT IS CLEAR THAT THE "DEVICE" TERM WAS ASSUMED TO REMAIN THE SAME FROM TEST TO TEST. SINCE THE VARIOUS COMPONENTS OF THE DILATO-METER ARE NOT NECESSARILY MADE OF THE SAME STOCK, HEATING RATES MAY VARY, AND SLIGHT RADIAL AND AXIAL THERMAL GRADIENTS ARE IN-EVITABLE, THE "DEVIATION" TERM IS DESIGNED TO COMBINE ALL OF THESE UNRESOLVED CONTRIBUTIONS BY THE INSTRUMENT. ITS ACTUAL VALUE IS OF NO SIGNIFICANCE AS LONG AS IT IS COMPARATIVELY SMALL AND, MOST OF ALL, REPRODUCIBLE.

TO MINIMIZE UNCERTAINTIES THE "DEVICE" TERM IS DETERMINED OR CROSS CHECKED FROM DIRECT OPTICAL DATA.

4.0 DISCUSSION OF RESULTS

THE DEVICE HAS BEEN USED TO MEASURE THE EXPANSION OF VARIOUS HIGH TEMPERATURE MATERIALS. A REPRESENTATIVE SAMPLE OF THE DATA OB-TAINED ON COMMERCIAL GRAPHITE STOCK IS SHOWN IN TABLE I. THE RESULTS ON THESE AND OTHER TEST RESULTS ON VARIOUS MATERIALS ARE FOUND TO BE IN EXCELLENT AGREEMENT WITH PUBLISHED FIGURES.

MODULATION METHOD FOR STUDYING THERMAL EXPANSION

Ya. A. Kraftmakher

Institute of Organic Chemistry

USSR Academy of Sciences, Novosibirsk

INTRODUCTION

At the present time methods for measuring dilatation of solids are known with a sensitivity of the order of 10^{-10} cm and even higher. However, difficulties in studying thermal expansion at high temperatures are associated with the instability of a sample rather than with the insufficient sensitivity. Therefore, one has to be content with the determination of average values of the thermal expansion coefficient within wide temperature intervals.

There are, however, two important physical phenomena in solids, the study of which requires the measurement of 'true' thermal expansion coefficients: point-defect formation and second-order phase transitions. In these cases relative changes in the volume or length of a sample are small whereas the thermal expansion coefficient changes substantially. For such measurements, it is convenient to employ the modulation method which involves oscillating the sample temperature about a mean value and recording corresponding changes in the sample length. Under these conditions the thermal expansion coefficient is directly measured. The use of the periodic temperature oscillations negates irregular changes of the sample length due to external disturbances or creep at high temperatures. In the modulation method only those changes of the sample length are registered which repeat reversibly with periodic changes of the temperature. Measurements can be performed with temperature oscillations of about 0.1 K. In principle even smaller temperature oscillations can be used but the actual resolution is limited by the variation of the mean temperature along the sample. At the present time, even measurements with a resolution of about 1 K at high temperatures are a great improvement over traditional methods.

Measurements by the modulation method can be performed with different ways of registering sample length oscillations including those
which yield the highest sensitivity.

Although the modulation method for studying thermal expansion
was proposed long ago (1) and there is already a review paper on
this subject (2) the method is still not widely known.

EXPERIMENTAL TECHNIQUE

Wire Samples

The sample is heated by a.c. current or by a current with both
d.c. and a.c. components (1). This results in the oscillation of
sample temperature around the mean value and in corresponding
periodic changes of sample length. One end of the sample is fixed
whilst the other is pulled by means of a load or a spring (Fig. 1a).
The free end of the sample is projected onto the entrance slit of
a photomultiplier with the aid of an optical system. The a.c.
component of the output voltage of the photomultiplier is proportional to the amplitude of the oscillations of the sample length. The
measuring system is calibrated under static conditions.

The amplitude of the oscillations of the sample temperature is
determined from the oscillatio ns of either the electrical resistance of the sample or its brightness (3). If the specific heat of
the sample is known, then the amplitude of the temperature oscillations can be easily calculated. When the sample is heated by a.c.,
the coefficient of thermal expansion is calculated from the expression

$$\alpha = 2mc\omega V/\ell PK,$$

where ℓ, m and c are the length, mass and specific heat of the sample;
P and ω are the power and frequency of the heating current; V is the
amplitude of the a.c. component at the output of the photomultiplier;
and K is the sensitivity of the photomultiplier to the dilatation
of the sample. The frequency of the temperature oscillations is
arranged to be sufficiently high that the phase shift between the
power and temperature oscillations is close to 90° (adiabatic regime).
Otherwise, the temperature oscillations should be calculated from
the formulae for the non-adiabatic regime (4).

When the sample is heated by a current with both d.c. and a.c.
components, the coefficient of thermal expansion is calculated

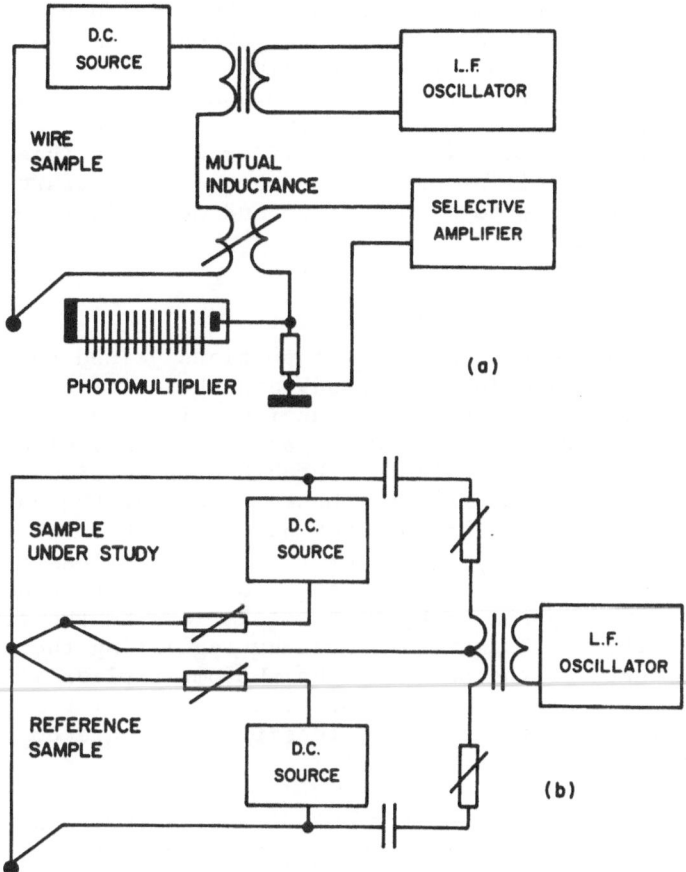

Fig. 1. (a) Modulation method of measuring the thermal expansion
             coefficient of wire samples (1).

        (b) Compensation method: oscillations of the length
             of the sample are compensated for by the anit-phase
             oscillations of a reference samples (5).

from the expression

$$\alpha = mc\omega V/2\ell KI_o U,$$

where U is the amplitude of the a.c. voltage across the sample.

In this case, a system can be employed whose balance is independent of the a.c. component of the heating current (Fig. 1a). The a.c. component at the out put of the photomultiplier is compensated for by means of a variable mutual inductance with the heating current being passed through its primary coil. The recording circuit incorporates a narrow-band amplifier which is tuned to the frequency of the oscillations of the sample temperature so that the system is insensitive to sample creep and mechanical perturbations.

Compensation Method

The wire sample consists of two portions joined together (Figure 1b): the sample under study and the reference sample with a known coefficient of thermal expansion. The two portions are heated by d.c. currents from two separate sources, with an a.c. current from a common oscillator through both portions. The polarity of the d.c. sources is chosen such that the current produces temperature oscillations 180° out of phase in the two portions of the sample. The amplitudues of these oscillations are adjusted so that the oscillations of the length of the portion under study are completely compensated by the oscillations of the length of the reference portion. This is achieved by regulating the a.c. current in the two parts of the sample. The photomultiplier is here only a zero indicator and the effect of variations in its sensitivity, as well as in the light source intensity, etc., are completely eliminated (5).

For calculating the coefficient of thermal expansion it is necessary to know the amplitudes of the temperature oscillations in the two parts of the sample. If the specific heat is known then it is easy to obtain the relation which holds when the oscillations in the two portions of the sample balance each other:

$$\alpha_1 I_{10} U_1 \ell_1 / m_1 c_1 = \alpha_2 I_{20} U_2 \ell_2 / m_2 c_2.$$

The subscripts 1 and 2 designate the studied and reference portions of the sample, respectively. The reference sample is at a constant mean temperature, and all the quantities except I_{10}, U_1, U_2, c_1, and α_1 are constants. Therefore

$$\alpha_1 = K U_2 c_1 / I_{10} U_1,$$

where K is a proportionality coefficient. Thus the measurements of the coefficient of thermal expansion are reduced to maintaining a zero amplitude of the oscillations of the length of the composite sample and to measuring the d.c. currents in, and the a.c. voltages

across, the two portions of the sample. The sensitivity of the
system is 10^{-7} cm.

Modulation measurements are best carried out with wire samples
but the determination of the temperature of such samples is a
difficult problem. It was shown (6) that the temperature of
current-heated wire samples can be determined from their thermal
noise. In the regime of periodic changes of the sample temperature, the noise thermometer (7) can be used for direct measurements of temperature derivatives of electrical resistance and heat
transfer of a sample – the quantities used for calculating the amplitude of temperature oscillations in the modulation technique.

Bulk Samples

Another version of the modulation method is used for comparatively bulky samples such as rods and foils. The temperature of
such samples and the amplitude of its oscillations are measured
with a thermocouple while the compensation of the oscillations of
the length is achieved by means of an electromechanical transducer
attached to the sample and controlled by a device sensing the
sample length, such as a small telephone transducer. A blade
mounted on the telephone membrane is illuminated by a light source
and its image projected at the entrance slit of a photodetector
(Figure 2). The output voltage of the photodetector is partly
compensated and then amplified by a d.c. amplifier. The output of
the amplifier is connected to the electromechanical transducer
with such a polarity that changes of the sample length are compensated by the displacements of the movable part of the transducer. Due to the high gain of the amplifier, almost complete compensation is automatically achieved (the error of the compensation
is less than 1%) and the sensitivity of the system is 10^{-6} cm. The
changes in the transducer current, which are proportional to the
changes of the length of the sample, are recorded (8).

For the elimination of cold-end effects, the temperature
oscillations are generated only in the central portion of the
sample. The sample is heated by a.c. mains current. A current
of the same frequency, but with its phase linearly varying with
time is superimposed on the mains current in the central portion
of the sample. As a result of the interaction of the two currents
the power dissipated in the central portion of the sample periodically varies and the temperature starts to oscillate around some
mean value.

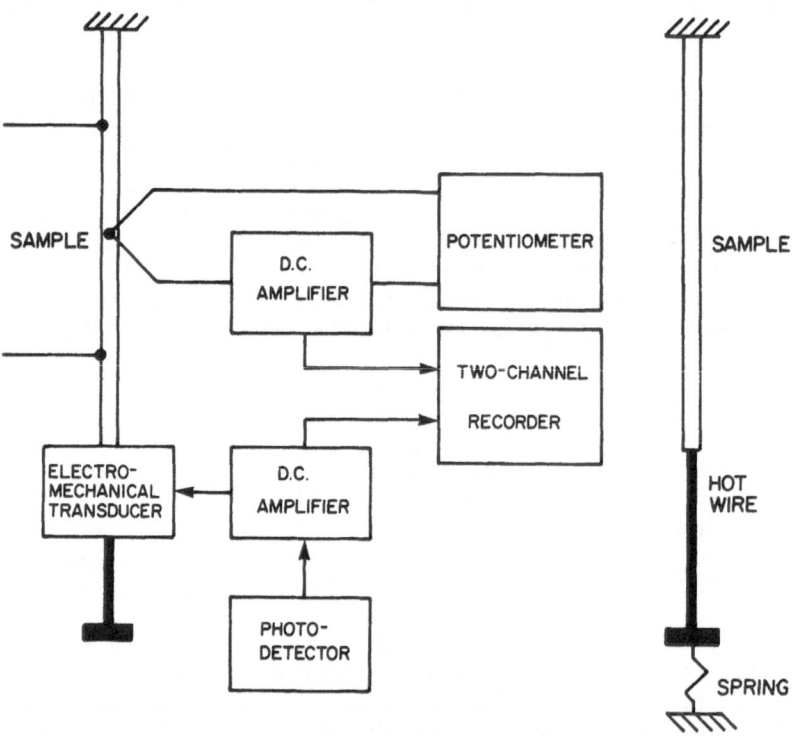

Fig. 2 Measurement of the thermal expansion coefficient of
        bulk samples: an electromechanical transducer or a
        wire heated by current is used for compensating the
        length oscillations (8).

Linearity of the electromechanical transducer used for compensation is very important. For this purpose, one can use the thermal
expansion of a thin wire heated by an electric current. One end of
the wire is fixed to the sample and the other is pulled by a spring.
A blade is mounted on the spring and its image is projected onto the
photodetector. Since the temperature oscillations of the sample are
small, only small changes of the temperature of the compensating wire
are required. With the mean temperature kept constant, the changes
in the temperature and length of the wire are proportional to the
changes in the current passing through it. A tungsten wire 0.05 mm
thick was used for compensation. At the mean temperature of 1500 K
the time response of such a transducer is quite sufficient and the
linearity is very good.

For determining the absolute values of the thermal expansion
coefficients, it is necessary to calibrate the electromechanical
transducer. For this purpose samples with a known thermal expansion

can be used. In some cases, (point-defect formation, phase transitions of the second order) it is sufficient to know the character of the temperature dependence of the thermal expansion coefficient, for which purpose only relative values are adequate.

DISCUSSION

Equilibrium Vacancies and Thermal Expansion of Metals

Measurements of the thermal expansion for studying the equilibrium vacancies in metals are widely used. There ar e two methods of measurement:

(1)   The separation of the vacancy contribution is made by an extrapolation from middle temperatures.

(2)   The dilatation of the sample and the lattice parameter are determined simultaneously. The equilibrium vacancy concentration is

$$c = 3(\Delta \ell / \ell - \Delta a / a)$$

The influence of the vacancy formation on the dilatation is relatively small but this effect becomes much larger if the thermal expansion coefficient is measured directly. With the modulation method is strong non-linear rise in the thermal expansion coefficient of platinum (9) and tungsten (10) was observed (Figure 3). The experimental data were approximated with the expression $\alpha = A + BT + (C/T^2)\exp(-E/kT)$, i.e. the temperature dependence of the thermal expansion coefficient for a vacancy-free crystal was suggested to be linear. The approximation by the least-squares method was performed using a computer. The dependence of the standard deviation on the chosen value of E has a minimum which indicates the most probable value of the vacancy formation energy. The values obtained in this way (Figure 4) are quite reasonab le although uncertainties in them are much larger than in values obtained from specific heat measurements (11, 12). To evaluate the equilibrium vacancy concentration, one may suppose that the vacancy volume is one half of the atomic volume (the vacancy formation is accompanied by a relaxation of the lattice). The rise in thermal expansion coefficient is

$$\Delta \alpha = (\gamma E A / k T^2) \exp(-E/kT),$$

where γ is the fraction of the atomic volume attributed to a vacancy, A is the preexponential factor in the expression for equilibrium vacancy concentration. Equilibrium vacancy concentrations calculated from the thermal expansion measurements (9, 10) were found to be smaller than those obtained from the specific heat measurements but again of the order of 1% at the melting points.

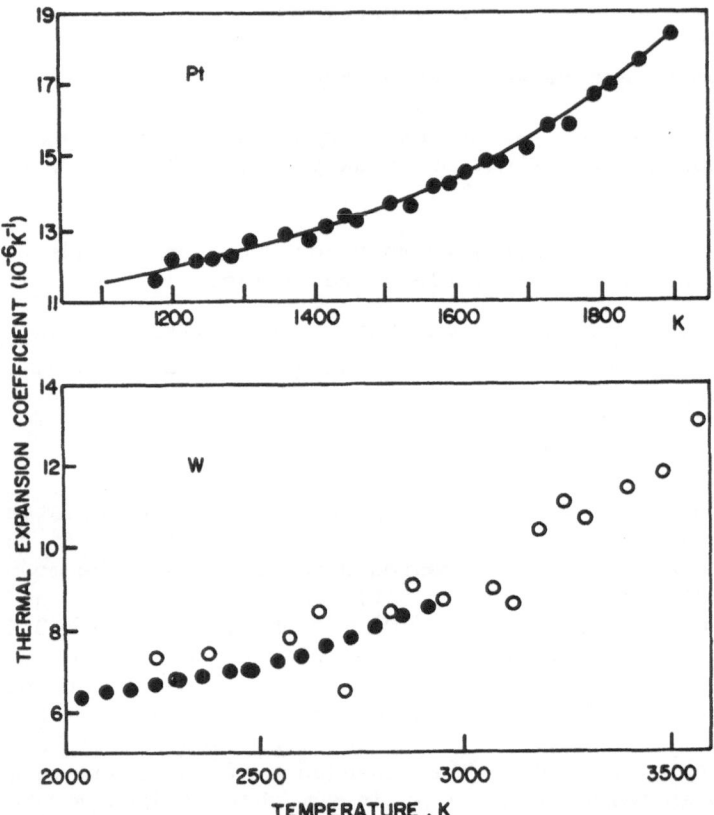

Fig. 3   Non-linear rise of the thermal expansion coefficient
of platinum and tungsten caused by vacancy formation

●   -   measurements by modulation method (9, 10),

○   -   calculations based on measured dilatation (13).

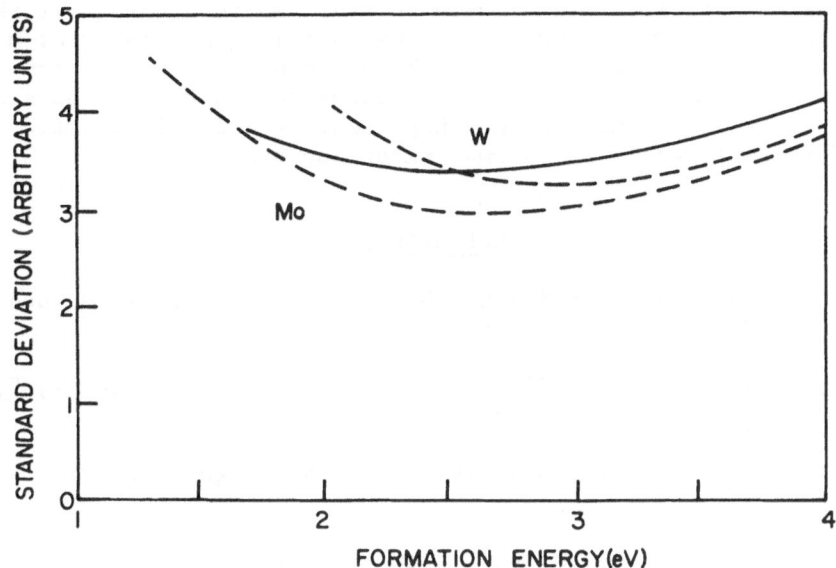

Fig. 4 Determination of the vacancy formation energy for tungsten and molybdenum using the thermal expansion data. ———— our results (10), ----- Petukhov and Chekhovskoi (13).

The conclusion on high equilibrium vacancy concentrations is not shared by many workers so this question remains disputable.

On comparing the data of different authors on thermal expansion a considerable discrepancy of results can be easily seen. A possible reason for these discrepancies has been already noted (11, 12). Since the sources and sinks for the vacancies are formed mainly by the internal imperfections of a crystal (voids, dislocations), the vacancy formation can partially take place without an increase of the total sample volume; the result of the measurements depends in this case on the character and density of the internal imperfections.

CONCLUSIONS

The modulation method for studying thermal expansion proved to be very convenient at high temperatures since it eliminates

difficulties specific to high-temperature measurements. It is
possible that this technique will be useful at middle or even at
low temperatures. Because of its high temperature resolution, the
method seems promising for studying anomalies in thermal expansion
near phase transitions. We may hope that it will be accepted as
one of reliable tools in thermal expansion studies.

REFERENCES

1. Ya. A. Kraftmakher and I.M. Cheremisina. Zh. Prikl. Mekhan.
    Tekhn. Fiz., No. 2 114, 1965.

2. Ya. A. Kraftmakher. High Temperatures - High Pressures, 5,
    645, 1973.

3. Ya. A. Kraftmakher. High Temperatures - High Pressures, 5,
    433, 1973.

4. A.A. Varchenko and Ya. A. Kraftmakher. Phys. stat. sol., 20a,
    387, 1973.

5. Ya. A. Kraftmakher. Zh. Prikl. Mekhan. Tekhn. Fiz., No. 4
    143, 1967.

6. Ya. A. Kraftmakher and A.G. Cherevko. Phys. stat. sol., 14a,
    K35, 1972.

7. Ya. A. Kraftmakher and A.G. Cherevko. Phys. stat. sol., 25a,
    691, 1974.

8. Ya. A. Kraftmakher and V.P. Nezhentsev. In: Fizika Tverdogo
    Tela i Termodinamika, Nauka, Novosibirsk, 1971, p. 233.

9. Ya. A. Kraftmakher. Fiz. Tverdogo Tela, 9, 1528, 1967.

10. Ya. A. Kraftmakher. Fiz. Tverdogo Tela, 14, 392, 1972.

11. Ya. A. Kraftmakher and P.G. Strelkov. In: Vacancies and
    Interstitials in Metals, North Holland, Amsterdam, 1970, p. 59.

12. Ya. A. Kraftmakher. J. Sci. Industr. Res., 32, 626, 1973.

13. V.A. Petukhov and V. Ya. Chekhovskoi. High Temperatures -
    High Pressures, 4, 671, 1972.

A SIMPLE DILATOMETER TECHNIQUE FOR

PHASE TRANSITION STUDY IN CRYSTALS

H.V. Tiwary, K.P. Sarbhai, and B.P. Adil

Physics Department, Ravishankar University
Physics Department, Government Science College
Raipur 492002 (M.P.) India

ABSTRACT

One of the techniques for the phase transition investigation
in crystals is the measurement of thermal expansion, since it is
correlated with several thermodynamic properties. A new simple
dilatometer is here reported. It is based on the principle of measuring the expansion of a needle-shaped material in terms of the
change in rotational angle of a reflecting surface with the help of
a spectrometer vernier scale. It has an overall sensitivity of
1 to 2 x 10^{-6}/°C. This dilatometer has been successfully employed
in the phase transition study of ferroelectric triglycine sulfate
and potassium nitrate. Our measurements of transition temperatures
and thermal expansion of these materials are in good agreement with
the values reported in the literature.

INTRODUCTION

The study of the physical properties of single crystals while
they pass through their transition temperature reveals interesting
behavior, enriching our understanding of solid state physics. There
is a close relationship between the nature of structural changes
and the variation in physical properties accompanying the phase
transition. Measurement of thermal expansion has been one of the
standard techniques for phase transition study in crystals. With
growing interest in this field it has been recognized that thermal
expansion measurements can be used to investigate the relation
between several thermodynamic properties. Further, it can provide
information regarding the order of phase transition. In a first-

order transition the energy, volume, and crystal structure change
discontinuously and the first derivative of the free energy shows
singularities. In a second-order transition, energy and volume
change continuously but the second derivative of the free energy
can have singularities. Thus the order of the transition can be
determined, using the volume expansion data. Thermal expansion
measurements will also indicate through thermal hysteresis whether
the phase transition is associated with any latent heat.

    A number of techniques are used for the measurement of thermal
expansion. X-ray diffractometry, electrical methods, and optical
techniques are usually adopted for thermal expansion studies, with
various modifications. Having worked with some of these techniques,
we have developed a much simpler but, at the same time, adequately
precise and informative technique.[1] The present paper deals with the
study of phase transitions in single crystals using this new technique.

EXPERIMENTAL

    The device consists of two parts, a spectrometer and a
dilatometer. The dilatometer shown in Fig. 1 was designed and fabricated by us in our department. It consists of a dilatometer table
(DT) which is mounted on a spectrometer in place of the prism table.
Its height can be adjusted and leveling done with the help of three
leveling screws. Along the coincident vertical axes of the spectrometer scale (SS) and the dilatometer, a flywheel (FW) loaded with
a hair spring (HS) is mounted with the help of two vertical upright
screws in the dilatometer table and the flywheel mount (FM). On the
flywheel stage, passing through the central slot in the axle, an
optical slide cover (SC) is mounted. This is also supported in the
back by a vertically mounted thin steel pin at the push point (PP).
This point is at a fixed distance of 2 mm from the axle of the
flywheel. The crystal is taken in needle form, 1 to 1.5 mm in
diameter and 1 to 3 cm in length. It is easy to shape the crystal
in this form or, in some cases, it can be grown in the needle form
itself. The crystal needle (CN) is kept horizontal in a miniature
furnace (MF), which is heated by two plane-parallel heating elements
(HH), 25 to 50 W each, covering the entire miniature furnace. There
are two supporting grooved screws on the two sides of the hollow
cylindrical (diam. 2 mm) miniature furnace to support the crystal
needle. For the measurement of the temperature of the crystal needle,
a small hole is provided at the center of one side of the miniature
furnace permitting the insertion of a chromel-alumel (32 gauge)
thermocouple junction which touches the crystal needle. This miniature furnace assembly is also mounted on the dilatometer table in
alignment with the push point of the flywheel and the intermediate
carrier needle as shown in Fig. 1. This carrier needle (N) is
preferably of low-thermal-expansion material. Its one pointed end

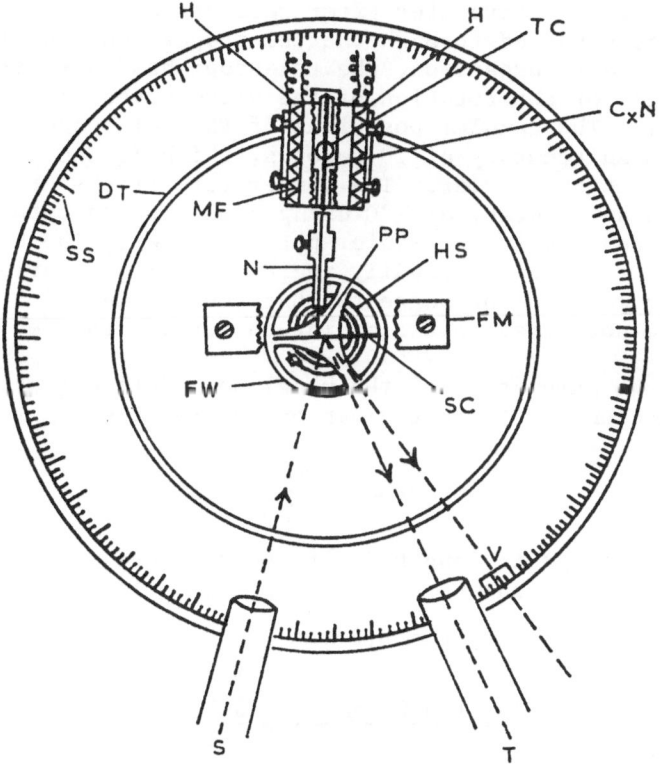

Fig. 1.   Schematic diagram of the simple dilatometer

touches the crystal needle and the other supports the flywheel push
point pin in the unwinding direction of the hair spring. The carrier
needle is encased in a heat-insulated tube where it is free to slide
when the expansion of the crystal needle is to be carried to the push
point. The mounting of these two needles is in a straight line,
normal to the plane of the slide cover (SC).

The expansion of the crystal needle causes a rotation of the
optical slide cover on the flywheel in proportion to the linear
change in length of the crystal needle. The incident ray from the
collimator of the spectrometer after reflection from the optical
slide cover is observed through the telescope. On rotation of the
slide cover, due to expansion, the telescope has to rotate through
twice the angle of the rotation of the slide cover to track the
reflected ray. The angular positions of the telescope can be
recorded with an accuracy of 10 seconds, which is the least count
of the spectrometer vernier. The temperature of the crystal is
measured with an accuracy of $\frac{1}{4}$°C using a calibrated chromel-alumel
thermocouple and precision vernier potentiometer. The heating of
the miniature furnace is steadily increased using a variac. Our
experience in measurement suggests that the problem of a temperature
gradient does not arise in the miniature furnace that we used.

Among the parameters of the usual formulation of the linear
thermal expansion coefficient given by the relation

$$\alpha = \frac{1}{\ell} \frac{\Delta \ell}{\Delta \ell} \tag{1}$$

$\Delta \ell$, i.e., the change in length due to temperature, is measured directly in the present technique in terms of the angle of rotation:

$$\Delta \ell = 0.2 \tan \Delta \theta$$

hence

$$\alpha = \frac{1}{\ell} \frac{0.2 \tan \Delta \theta}{\Delta T} \frac{1}{\ell} \frac{0.2 \Delta \theta}{\Delta T} \tag{2}$$

(for small angle of rotation). Here $\Delta \theta$ is the angle of rotation of
the optical slide cover due to expansion of the crystal needle whose
temperature is raised by ΔT °C.

The overall sensitivity of our spectro-dilatometer technique
for thermal expansion measurement is about 1 to 2 x 10^{-6}/°C. The
overall estimated accuracy is about 6 to 7%.

RESULTS AND DISCUSSION

The spectro-dilatometer technique has been successfully and
conveniently employed to measure the phase transition temperatures

of certain needle-shaped ferroelectric crystals by studying the variation of their respective thermal expansion coefficients with temperature. The rate of heating was kept uniform and slow, viz., $\frac{1}{2}$°C per minute. The two ferroelectric crystals chosen were potassium nitrate and triglycine sulfate (TGS), the former exhibiting phase transition at 125°C and the latter between 46°C and 49°C, as reported in the literature. The range of temperature covered by us extends from room temperature to beyond the transition temperatures. A number of needle-shaped potassium nitrate specimens were grown for study and their thermal expansion coefficients measured along the c-direction. As shown in Fig. 2, the value of α_c varies linearly up to about 110°C and then increases rapidly, resulting in the cracking of the crystal itself near the transition region, viz., at 120°C. The value of α_c for potassium nitrate at 50°C was found to be 300 x 10^{-6}/°C, the transition having been ascertained to be of the first order as was obvious from cracks developed in the specimen and the sudden variation of α_c around the transition temperature.

Measurement of thermal expansion coefficients for the various specimens of TGS crystals along the a-direction yields the value 34.8 x 10^{-6}/°C at 52°C. The crystals showed a negative anomaly at the transition temperature of 46°C, as is evident from a sharp dip (Fig. 3). The transition in this case was conclusively of the second order. The transition temperatures and the values of the coefficients of linear thermal expansion as obtained from our measurements are in fair agreement with values in the literature[2,3] for these crystals.

The simple formula used to calculate the thermal expansion coefficient owes its validity to a few essential requirements such as (i) normalcy of the carrier needle to the optical slide cover in all positions of the latter, (ii) small angle of rotation of the slide cover resulting from the expansion of the specimen, (iii) uniformity of temperature within the miniature furnace, (iv) coincidence of the axes of the dilatometer and the spectrometer, and (v) small thickness of the slide cover. During the measurements for smaller angles of rotation the errors introduced will be negligible. For larger variation (10 to 15°) in the angle of rotation, the normalcy of the carrier needle to the optical slide cover cannot be maintained. For this the corrected value of the angle can be estimated from relation (3), which can be easily derived, the corrected value (θ') being a function of the observed value θ:

$$\theta' = \tan^{-1}(\sec \theta \cdot \tan \theta) \qquad (3)$$

The error due to noncoincidence of the axes can be eliminated by measuring angles when the slide cover is rotated in turn in both directions, clockwise and counterclockwise. If, however, the slide cover has a thickness t, the measured value of $\Delta\theta$ will always be greater than the actual value. In our measurements, the error due to this effect was only about 0.15%. However, the corrected value

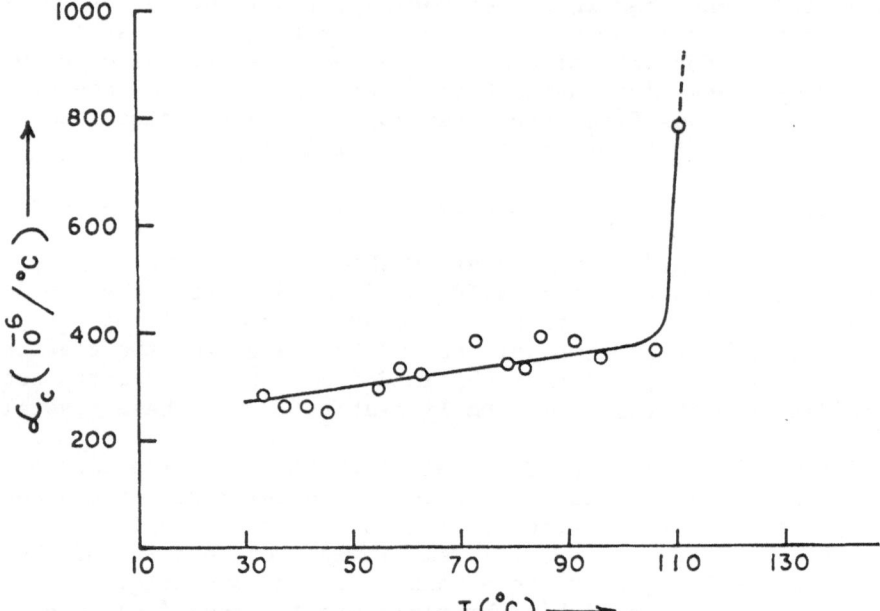

Fig. 2. Temperature variation of the thermal expansion
coefficient of KNO_3 along the c-axis.

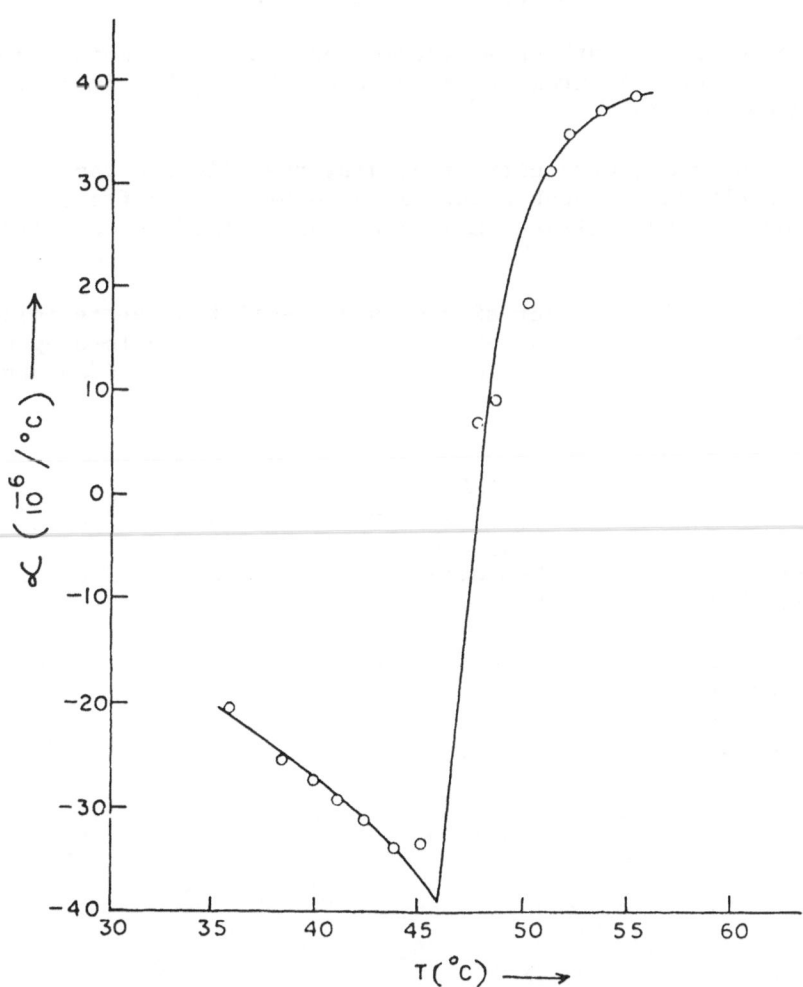

Fig. 3. Temperature variation of the thermal expansion coefficient of TGS.

of $\Delta\theta$ can be calculated from the following relation:

$$\Delta\theta' = \cos^{-1}\frac{r\cos\theta + t}{r - t\cos\Delta\theta} \qquad (4)$$

where r is the radius of the spectrometer scale.

The expansion of the miniature furnace is not going to cause any error in our measurements as care has been taken in design to compensate for it.

By using a quartz carrier needle, the slight expansion of the needle itself at the push point can also be made ineffective as quartz has a very small thermal expansion coefficient compared to other materials.

We conclude that our simple spectro-dilatometer technique described for thermal expansion measurements of needle-shaped specimens can be fruitfully employed for phase transition studies in solid materials with adequate precision.

REFERENCES

1. H.V. Tiwary, M.P. Gupta, and C.P. Pandey, Proceedings of Solid State Physics Symposium (BARC), Ahmedabad, Dec. 27-31, 1976, Vol. 19c, p. 498.

2. V. Kawabe, T. Yanagi, and S. Swada, J. Phys. Soc. Japan 20:2059 (1965).

3. I. Shibuya and S. Hoshino, Jap. J. Appl. Phys. 1:249 (1962).

A HIGH PRECISION PRIEST INTERFEROMETER

W. A. Plummer
H. E. Hagy

Research and Development Laboratory
Corning Glass Works
Corning, N. Y. 14830

ABSTRACT

A novel Priest-type interferometer has been built to measure small differences in expansion of fused silica. The interferometer block consists of a 3-cm. diameter cylinder, 10 cm in height, wrung onto a shorter cylinder. The sample is supported in a 0.64-cm off-axis hole. Two 5-μm diameter silica fibers and the upper curved surface of the sample provide three point contact with the semi-circular upper flat. Temperature control is accomplished by circulating fluid from a constant temperature bath through the walls of a double-walled chamber. The fringe separation is measured with an automatic counter. The apparatus is appropriate for measuring the expansion of any material having an expansion close to that of fused silica. Absolute expansions can be measured by appropriate calibration of the interferometer using specimens of known expansion.

INTRODUCTION

Fused silica is manufactured by Corning by the flame hydrolysis of $SiCl_4$. A similar process using a combination of $SiCl_4$ and $TiCl_4$ is used for the manufacture of an ultralow expansion titanium silicate glass, Code 7971.

A Fizeau interferometer[1,2] was developed to measure the
temperature of zero-expansivity, which occurs near room
temperature. At temperatures below zero the expansivity
of Code 7971 becomes highly negative. The expansivity
of fused silica on the other hand, becomes smaller with
temperature, reaching zero at about -120°C. The expansion characteristics of high silica glass depends not
only on its composition but also the thermal history it
has undergone.[3,4,5] Variations in expansion due either
to the manufacturing process or to subsequent operations
must be controlled within tight limits for reliable performance in critical applications. Therefore a method
for measuring small expansion differences was required.

METHOD

A Priest[6] (or Dental) interferometer was chosen.
This is a differential technique in which the expansion
of the sample is measured relative to that of a reference material. As the sample material expands or contracts relative to the reference material, the angle between the two optical flats changes. This change in
angle causes the spacing between the fringes to increase
or decrease. The following equation expresses this relationship:

$$(\Delta L/L)_M = (\Delta L/L)_S - (\Delta L/L)_R = (\lambda/2)\,(D/L_o)\,(N_2 - N_1) \quad (1)$$

where $(\Delta L/L)_M$ = measured differential expansion

$(\Delta L/L)_S$ = sample expansion

$(\Delta L/L)_R$ = reference expansion

λ = wavelength of light used

D = perpendicular distance between the
specimen bearing point and the line
joining the reference material bearing points

N_1, N_2 = number of fringes per unit distance at temperatures T_1 and T_2

APPARATUS

The Priest interferometer consists of a cylinder 3 cm in diameter and 10 cm in length with its top surface polished to the optical flatness of 1/20th of a wave. The cylinder is wrung onto a shorter cylinder which provides support for the sample. The sample rests in an off-axis hole 0.64 cm diameter. The tip of the sample and two 0.13 mm diameter silica fibers define the three bearing points for the top optical flat. Three horizontal thermocouple holes serve to monitor the axial temperature distribution.

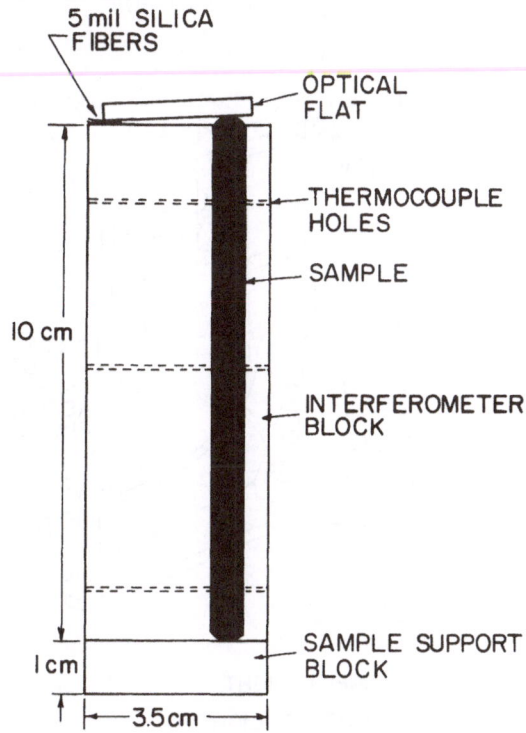

Fig. 1 Schematic of Priest Interferometer

A Gaertner Model I-1121 pulfrich viewer[7] was modified by Gaertner to permit simultaneous photoelectric
fringe detection and viewing. The combined collimator
and telescope focus the fringes onto a mirror with a
slit. Part of the light passes through the slit and is
detected by the sensing phototube. The remainder of the
light is received by the reference phototube after being
attenuated by a neutral density filter. The use of two
phototubes in a bridge type circuit corrects for variations in intensity of the light source or other changes
which affect both tubes. The output obtained is proportional to the fringe opacity.

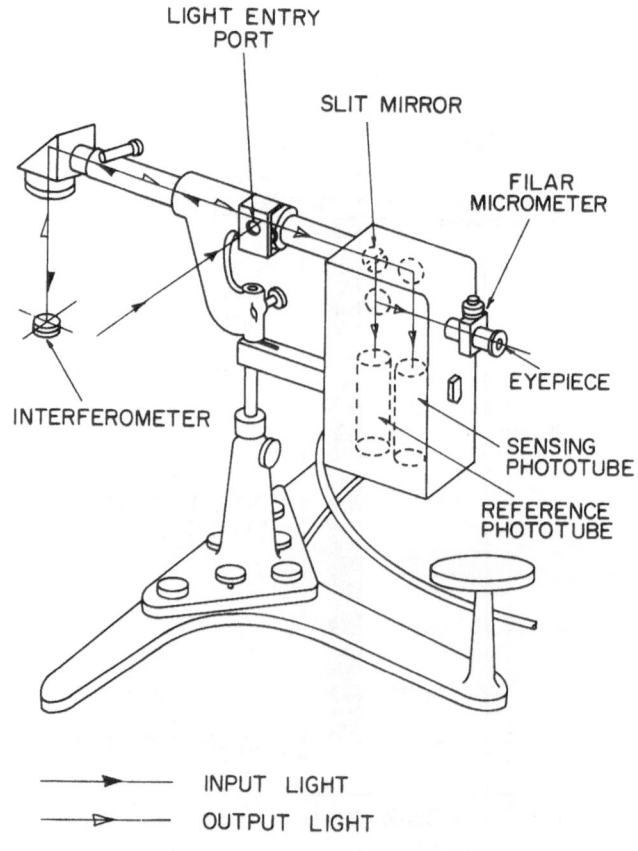

Fig. 2 Viewing Assembly

When used for Fizeau interferometry the counter records the movement of fringes past the slit. In the Priest interferometer, however, it is necessary to move the specimen past the slit. To accomplish this the sample chamber rests on a moveable table. The table, 10 x 16 cm, has a micrometer screw with one turn equal to 0.127 cm (.050"). The movement of the table is measured using a Collins SS105 LVDT. The coil LVDT is mounted on the stationary section and the core on the moveable section. The LVDT output is attenuated by a 10 turn precision potentiometer and measured by a recorder.

An L & N Speedomax G X-Y recorder is used to measure and record the fringe spacings. This recorder has an adjustable zero (± 50 mv) and adjustable range (1 - 100 mv). The phototube output is fed to the Y-axis, that from the LVDT to the X-axis.

A double walled aluminum cylinder, 7.6 cm I.D., 14 cm O.D. serves as the temperature controlled chamber for the sample. The cylinder is insulated using a 3-cm thick sponge foam. A 5-cm thick block of polystyrene insulates the cylinder from the table. A Model FE Haake temperature-controlled bath circulates water through the walls of the cylinder. A copper cylinder, 4.5 cm I.D. and 6.4 cm O.D., is inserted around the sample to improve the temperature uniformity without sacrificing ease of sample set-up.

The light source is a Spectra-Physics Model 119 single frequency He-Ne gas laser. A mercury arc lamp or equivalent light sources are satisfactory alternatives for generating fringes between the small plate/block gap.

SAMPLE PREPARATION

Samples are core drilled and ground to 0.632 cm diameter. This diameter permits free vertical movement but severely restricts any lateral motion of the sample. Restricting the axial tilt to 0.002 cm limits the cosine error to 0.02 ppm. Spherical surfaces are ground on each end and the length slowly adjusted to within 0.05 mm of the desired length. This will give about 8 fringes per mm. The final length is carefully adjusted to give

between $\frac{1}{2}$ and 1 fringe per mm. This corresponds to a
tolerance of 0.005 ± .002 mm in the desired length and
represents a major limitation of the method.

LVDT CALIBRATION

A Collins Model SS105 LVDT is used to measure the
motion of the table as the fringes are moved past the
fringe counter. The sensitivity of this model is in
excess of 100 mv/mm, which is much greater than needed
for our purposes. The output of the LVDT is reduced by
a voltage divider network using a 10 turn precision
potentiometer. A setting on the potentiometer was chosen
to give about 4 mv/mm displacement. Thus, for 25 mm
total displacement the full 100 mv range of the recorder
can be utilized. The relative core/coil position was
adjusted to give close to zero output with the screw in
its midrange position. A series of readings were taken
every turn of the micrometer screw (2.54 mm). These
values are shown in Table I. The output of the LVDT is
not linear over the desired range. The sensitivity is
lower by 10% at 12 mm displacement. The quantity
$(L - L_o)$/mv was plotted vs the mv output and an empirical equation found for expressing the displacement as a
function of the LVDT output. This expression for the
displacement in millimeters is:

$$(L - L_o) = 0.2507 \ (mv) + 6.71 \times 10^{-6} \ (mv)^3 \qquad (2)$$

The precision with which the calibration constant is
known is 0.0011 mm/mv. Since the fringe spacing is
measured over the same range of outputs, the effect of
any error in the value of this constant should be reduced. The error in the difference in fringe spacing
for a typical case is reduced by a factor of five.

MEASUREMENT OF FRINGE SPACING

Fringes can be measured over a range of about 25 mm,
with 2 to 20 fringes being the desired number. A typical trace is shown in Fig. 3 where the distance between
fringes is about 4 mm, thus providing 5 light and 5 dark
fringes whose positions can be determined. To locate

the peak maxima (or minima), the recorder is set on the
5 or 10 mv range which provides ample sensitivity. The
center of the peak is located by taking equal intensities on either side of the peak, reading and averaging
the LVDT voltages at those positions as determined from
the recorder. This voltage is converted to a displacement by means of equation (2).

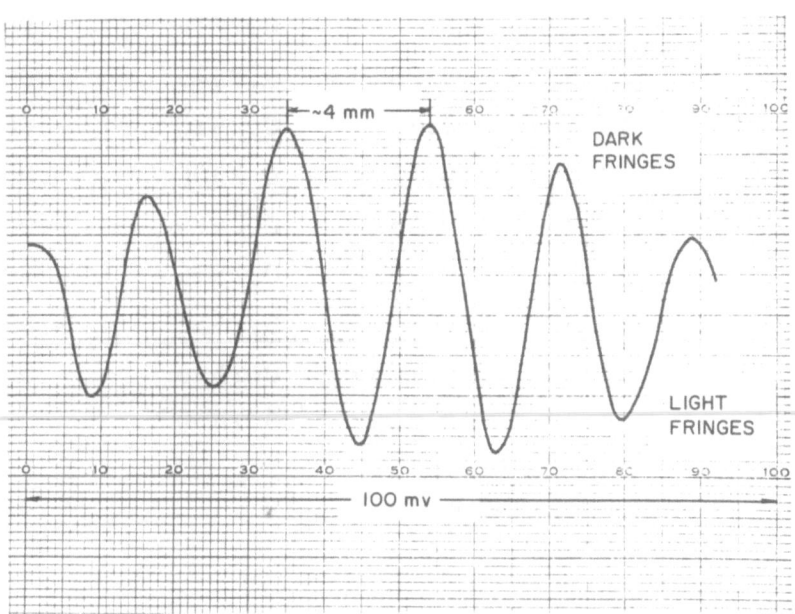

Fig. 3   Recording trace of fringes

The precision with which fringe spacings can be determined is influenced by the number, width and spacing
of the fringes. For a spacing of 4 mm between fringes
the position of 6 dark and 6 light fringes can be determined. A single fringe can be measured to a precision
(1 σ) of 0.004 mm. For expressing the uncertainty in
fringe spacing the standard error of the mean (SEM) is
useful which in the example chosen is 0.035 mm. This is
equivalent to an error of 0.0021 mm^{-1} in the spacing
expressed in fringes/mm. The corresponding error in
$\Delta L/L$ will be 0.14 ppm. Similar analyses have been carried out for other fringe spacings with the conclusion
that the expansion error is essentially independent of
fringe spacing. This is illustrated in Table II.

Table I

LVDT CALIBRATION

Displacement $L-L_o$, mm	LVDT Output mv	$(L-L_o)/mv$ meas'd	$(L-L_o)/mv$ calc'd	$(L-L_o)/mv$ calc'd-meas'd
-11.133	-42.32	0.2631	0.2627	-.0004
- 8.598	-33.22	0.2587	0.2614	+.0027
- 6.053	-23.83	0.2540	0.2545	+.0005
- 3.513	-13.96	0.2516	0.2520	+.0004
- .973	- 3.88	0.2508	0.2508	.0000
1.567	6.24	0.2511	0.2510	-.0001
4.107	16.29	0.2521	0.2525	+.0004
6.647	25.99	0.2558	0.2552	-.0006
9.187	35.27	0.2605	0.2590	-.0015
11.727	44.39	0.2644	0.2639	-.0005

$$\sigma = 0.0011 \text{ mm/mv}$$

Table II

Analysis of Measurement Precision

Fringe Spacing mm	No. of fringes	σ mm	SEM mm	$\Delta(1/\bar{X})$ mm	$\Delta(\Delta L/L)$ ppm
1.0	20	0.010	0.002	0.002	0.16
2.0	20	0.030	0.008	0.002	0.16
3.0	16	0.060	0.015	0.0017	0.12
4.0	12	0.100	0.030	0.0019	0.13
5.0	10	0.160	0.050	0.002	0.16
6.0	8	0.230	0.080	0.002	0.16
7.0	5	0.325	0.145	0.003	0.20

RESULTS

Four specimens of Code 7940 were core drilled from the same block which provided the material for the interferometer. Data were obtained on three of the specimens. For two of the samples increasing temperatures caused a small decrease in distance between fringes and the other showed an increase in fringe spacing. By measuring the absolute lengths we ascertained that the third specimen had been polished too short so that the top flat angled downward from the fiber supports to the sample. This negative angle will be reduced if the specimen expands and the distance between fringes will increase. If the angle is positive, expansion of the specimen will increase the angle which decreases the fringe spacing. All three samples exhibit an expansion relative to the block of 0.006 ppm/°C. The results of these experiments are illustrated in Fig. 4. The experience with the short sample illustrates the difficulty of sample preparation and serves as a precaution for future measurements. However, because the one sample was short several alternative explanations for the positive expansion difference were eliminated.

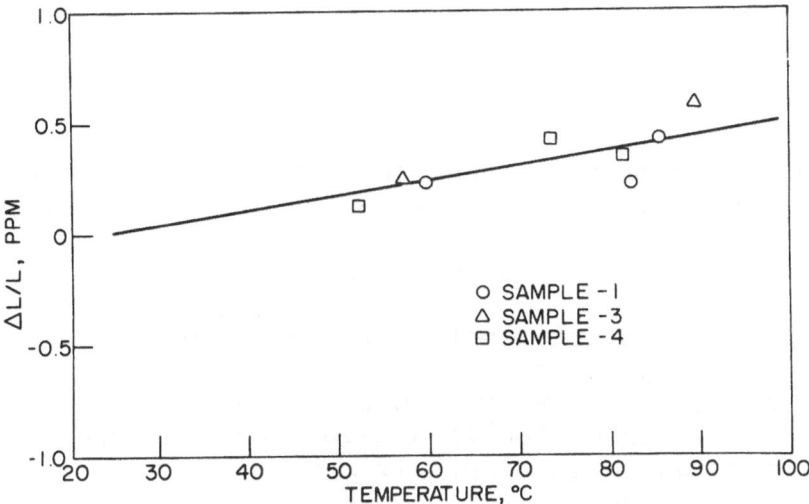

Fig. 4. Measured differential expansion of samples from the interferometer block of Code 7940

Two specimens of Code 7971 were also measured.
This material has a "high" negative expansion with respect to Code 7440. The angle between the flats will
decrease giving an increase in the fringe spacing. A
temperature change of 60°C will show a differential expansion of -33 ppm. A fringe spacing of 1.5 mm at 25°C
will increase to 6 mm at 85°C. If the spacing is 1.7 mm
at 25° it will be 11 mm at 85°C. Fringe spacings of
this magnitude or larger present difficulties in determining fringe position and serve to further illustrate
the problems in sample preparation.

The differential expansion for these two specimens
of Code 7971 are given in Tables III & IV and depicted
in Figs. 5 and 6. These values are relative to the interferometer block expansion. The accuracy of these
measurements depends upon the accuracy with which the
lever distance "D" can be determined. The two specimens
would normally be expected to have the same temperature
dependence. However, the 337-402 sample had a special
heat treatment which has resulted in a more negative expansion. At 25°C this sample is more positive by 0.01
ppm/°C. than the 347-132E sample. Jacobs[8] value at 25°C
for 337-402 was -0.008 ppm/°C. Ultrasonic measurements
on the 347-132 boule gave -0.020 ppm/°C. for this specimen. The difference is very close to that determined
in these experiments.

The absolute expansion values obtained by Jacobs
can be combined with our measured values using Equation
(1) to determine the expansion of the interferometer
block. The results so obtained are shown in Table V.

CONCLUSIONS

A Priest interferometer of a novel design has been
shown to be capable of measuring expansion coefficients
to a precision of 5×10^{-9}/°C over a 30° temperature interval. This precision is attained by employing a 10 cm
long sample and utilizing a fringe counter to measure
changes in fringe separation. Measurements have been
made on samples of fused silica and titanium silicate
glass and the expansion of the interferometer block has

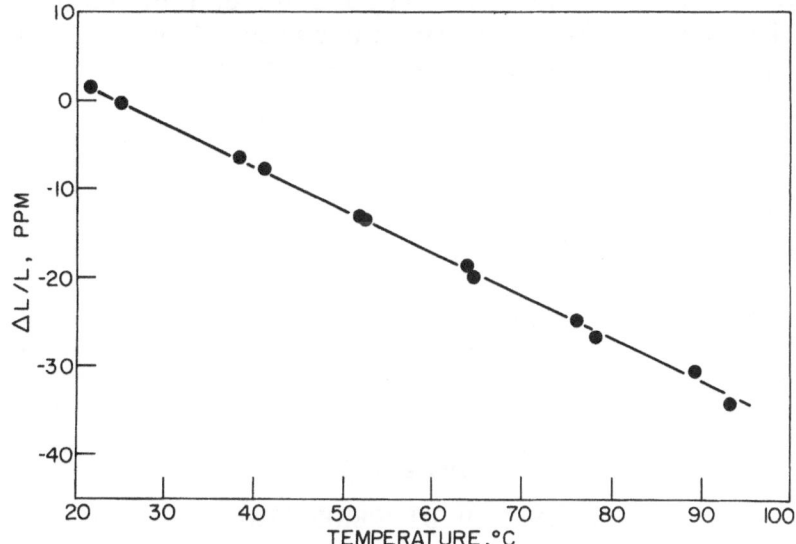

Fig. 5. Measured differential expansion of
Code 7971, boule 347-132 (E)

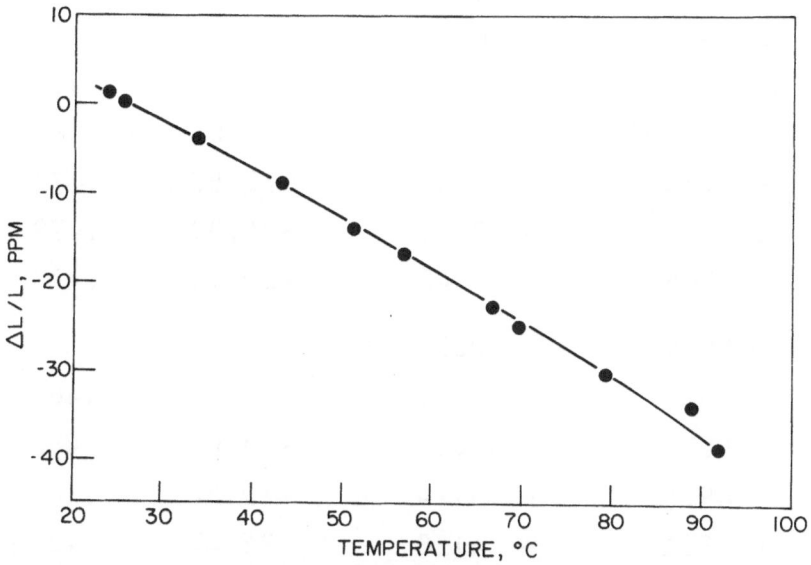

Fig. 6. Measured differential expansion of
Code 7971, boule 337-402

been calculated. The interferometer can be valuable in
measuring the variations in expansion of fused silica
arising from differences in composition or in the thermal history.

Table III
MEASURED DIFFERENTIAL
THERMAL EXPANSION
CODE 7971
347-132E

TEMP °C	1/X Fringes/mm	$\wedge$ (1/X) Fringes/mm	$\Delta L/L$ ppm	$\Delta L/L (T-25)$ $10^{-6}/°C$
21.6	0.9041	+ .0254	+ 1.55	- .456
38.3	0.7782	- .1005	- 6.15	- .462
51.8	0.6658	- .2129	-13.03	- .486
63.8	0.5764	- .3023	-18.50	- .477
75.8	0.4812	- .3975	-24.33	- .479
89.1	0.3845	- .4942	-30.24	- .472
25.2	0.8071	+ .0018	-	-
41.0	0.6930	- .1159	- 7.38	- .461
52.0	0.6039	- .2050	-13.05	- .483
64.6	0.4975	- .3114	-19.82	- .500
78.2	0.3931	- .4158	-26.47	- .498
92.8	0.2744	- .5345	-34.02	- .502

Table IV
MEASURED DIFFERENTIAL
THERMAL EXPANSION
CODE 7971
337-402
SUB-MASTER STANDARD

TEMP. °C	1/X Fringes/mm	Δ (1/X) Fringes/mm	ΔL/L ppm	ΔL/L/(T-25) 10^{-6}/°C
23.3	0.7424	0.0137	0.96	-
50.8	0.5263	- .2024	-14.14	- .548
69.6	0.3759	- .3528	-24.66	- .553
88.5	0.2493	- .4794	-33.51	- .528
25.0	0.4544	-	-	-
33.7	0.5157	0.0613	- 3.98	-0.457
42.7	0.5896	0.1352	- 8.78	- .496
56.5	0.7112	0.2568	-16.67	- .529
66.6	0.8019	0.3475	-22.55	- .542
78.9	0.9149	0.4605	-29.89	- .554
91.6	1.0449	0.5905	-38.32	- .575

Table V
Expansion of Interferometer

Temp. °C.	$\Delta L/L(T-25)$, $10^{-6}/°C$		
	337-402	Measured	Interferometer
30	- 0.005	- 0.445	+ 0.440
40	0.000	- 0.486	0.486
50	0.005	- 0.514	0.519
60	0.010	- 0.535	0.545
70	0.015	- 0.555	0.565
80	0.020	- 0.560	0.580
90	0.025	- 0.564	0.589

REFERENCES

[1] W. A. Plummer and H. E. Hagy, Applied Optics 7 825 (1968).

[2] W. A. Plummer, AIP Conference Proceedings No. 3, Thermal Expansion-1971, 36, American Institute of Physics (1972).

[3] R. Bruckner, Glastechn. Ber. 37 413, 459, 500 (1964).

[4] J. Oishi and T. Kimura, Metrologia 5 50 (1969).

[5] T. A. Hahn and R. K. Kirby, AIP Conference Proceedings No. 3, Thermal Expansion-1971, 13, American Institute of Physics (1972).

[6] ASTM Method E289 "Linear Thermal Expansion of Rigid Solids with Interferometry".

[7] C. J. Parker and W. A. Popov, Applied Optics 10 2137 (1971).

[8] S. Jacobs, Optical Science Center, Univ. of Ariz. private communication.

EXPANSIVITY OF SILICON 20 - 500°C

R.B. Roberts

National Measurement Laboratory, CSIRO

Sydney, Australia 2008

ABSTRACT

A program is in progress to extend the highly accurate data for the thermal expansion of silicon[1] to at least 500°C by means of a polarization interferometer.[2] In this instrument a laser beam is split and recombined after being reflected from (effectively) both ends of a polished sample. For a 15 mm sample, length changes of 3 nm can be resolved. The accuracy of the thermal expansivity measurement is thus limited by the resolution and accuracy of the thermometer.

The furnace has recently been modified to improve temperature uniformity and the degree to which the thermometer is bonded to the sample. At the time of writing only a few measurements have been made but they agree with Lyon et al.[1] within $0.01 \times 10^{-6}/°C$ at room temperature.

INTRODUCTION

Many changes have been made to the furnace tube and temperature measurement systems since the first description of the apparatus.[2] In its present form it is capable of resolving changes of length of about 3 nm in a sample of 15 mm length in the region from 20 to 500°C. Measurements on the expansivity of semiconductor-grade silicon have been made but the scatter of the data still exceeds 1%, so the instrument is still under development. However the limitation is probably better expressed as being $\pm 4 \times 10^{-8}/°C$ so the performance should be relatively better on a material with a larger expansion coefficient.

FURNACE

The furnace tube was machined from solid copper, 25 mm O.D. X 200 mm long with 3 mm walls and a 20 mm platform near the centre. The heating elements were wound in three separate sections from nichrome wire covered in glass fibre sheathing. Differential thermocouples were incorporated to measure the temperature uniformity and aid in its control. The end heaters can always be adjusted to within 0.1 °C of the middle (sample) temperature. The copper tube is supported by four 2 mm O.D. X 40 mm long stainless steel tubes.

THERMOMETRY

Most of the development work has been concerned with getting the thermometers to the same temperature as the sample. The leads are now brought up the outside of the copper tube in close contact with the heaters and halfway down the inside of the copper tube to give an effective 'immersion length' of 300 mm. The effect of conduction errors in the leads is further reduced by making the last 100 mm of fine (0.1 mm) platinum wire. The thermometer is now a Rosemount type 2050. It fits into a hole in the active part of the sample. Provision is also made for a Pt/Pt10%Rh thermocouple as before.

RESULTS

The data on a piece of silicon cut from the core of the sample used by Norton et al.[3] for their room temperature survey are given in Fig. 1. They are compared with the low temperature data from Lyon et al.[1] and the higher temperature data of Plummer.[4] The present data were calculated on direct measurements of $\Delta L/\Delta T$ whereas all the others were smoothed before plotting. On such a basis it can be said that all measurements agree within $\pm 3 \times 10^{-8}/°C$, the present measurements extending the observed region to about 500°C.

INTERFEROMETER POSSIBILITY

With the present instrument the performance of the optics seems to be limited by the shear compensation device or perhaps sample steadiness. The incorporation of a polarizing beam splitter and quarter wave plate as in Fig. 2 could be an improvement. For one thing a longer sample could be used more easily. For another it would be easier to use a long focal length lens which should then be even more tolerant of changes in the sample orientation than is the case at present.

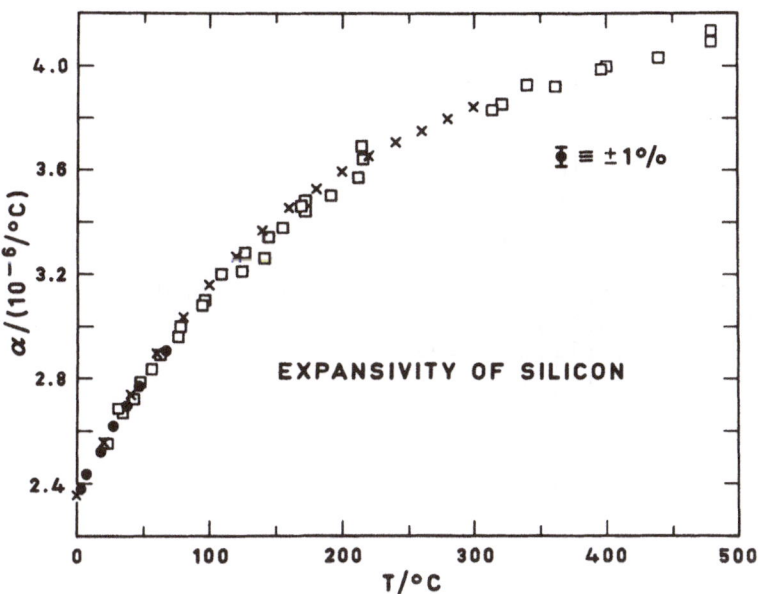

Figure 1

Thermal expansivity of silicon

● Lyon $et_4al.$ [1]
X Plummer
□ present results

It has been found that the nonlinearity of response arising from the difference in transmission of the recombining beam elements was overstated. With further experience at setting up the beam the author is now unable to detect the nonlinearity.

CONCLUSION

Further minor development work is still required before the instrument will produce the $\pm 1 \times 10^{-8}/°C$ accuracy in expansion coefficient that was intended.

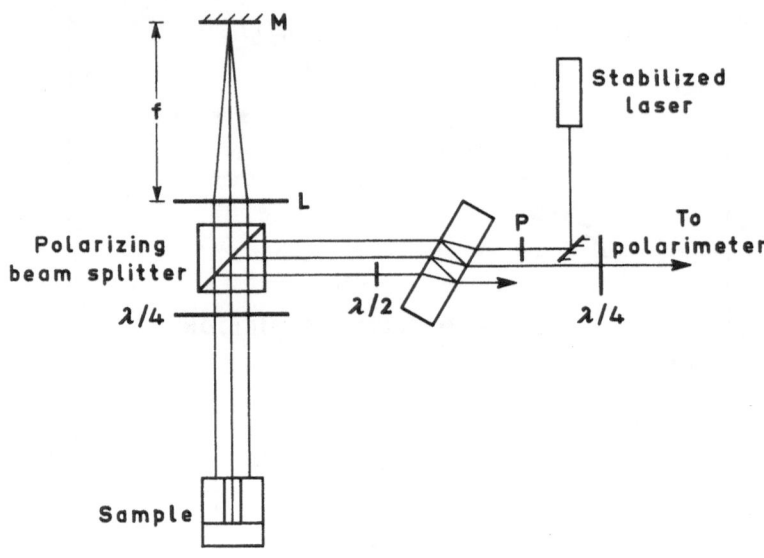

Figure 2

Possible improved form of interferometer

REFERENCES

[1] K.G. Lyon, G.L. Salinger, C.A. Swenson and G.K. White, J. Appl. Phys. 48, 865-8 (1977).

[2] R.B. Roberts, J. Phys. E 8, 600-2 (1975).

[3] M.A. Norton, J.W. Berthold III, S.F. Jacobs, and W.A. Plummer, J. Appl. Phys. 47, 1683 (1976).

[4] W.A. Plummer, private commun.

THERMAL EXPANSION OF SINGLE CRYSTAL SAPPHIRE FROM 293 TO 2000 K

STANDARD REFERENCE MATERIAL 732

Thomas A. Hahn

National Bureau of Standards

Washington, D. C. 20234

ABSTRACT

This single crystal sapphire is the fifth Standard Reference Material (SRM) to be certified for thermal expansion by the National Bureau of Standards. Expansion measurements to 1000 K on three samples indicate that the stock of material is of consistent quality suitable for certification. In the temperature range from 293 to 1000 K measurements were made using an interferometer apparatus resulting in

$$\frac{L_{1000} - L_{293}}{L_{293}} = 5500 \times 10^{-6}$$

with a standard deviation of 5×10^{-6}. Above 1000 K length measurements were made using twin tele-microscopes with the sample heated in a tungsten mesh furnace. From 293 to 2000 K

$$\frac{L_{2000} - L_{293}}{L_{293}} = 15700 \times 10^{-6}$$

with a standard deviation of 25×10^{-6}. Measurements were made parallel to the rod axis which is oriented 59° from the c-axis of the crystal. Smoothed values of the expansion and coefficients of expansion were obtained by fitting the expansion with a variation of Grüneisen's equation

$$\frac{L_T - L_{293}}{L_{293}} = \frac{1}{3a} \quad \frac{E\left(\frac{\Theta}{T}\right)}{Q_0 - kE\left(\frac{\Theta}{T}\right)} + \frac{1-a}{a}$$

Final values of the four parameters were obtained by minimizing the standard deviation in a least-squares analysis. The results of this study are generally in good agreement with expansion values found in the literature. Major differences, however, are observed in the comparison with x-ray diffraction studies and with extrapolated data appearing in the literature. No systematic deviations of the data from the final equation were observed. A comparison of the results of this study with data found in the literature will be presented.

INTRODUCTION

This report is part of a program by the National Bureau of Standards to certify a number of thermal expansion Standard Reference Materials (SRM) to aid in the calibration and verification of thermal expansion measurements. Materials that have been certified and their temperature ranges are: SRM-736 Copper from 20 to 800 K, SRM-739 Fused Silica from 80 to 1000 K, SRM-731 Borosilicate Glass from 80 to 680 K, SRM-737 Tungsten from 293 to 1800 K, and SRM-732 Single Crystal Sapphire from 293 to 2000 K (1). Sapphire and tungsten extend the calibration capabilities to the technologically important high temperature range where data uncertainties can be large due to the difficult measurement conditions.

Single crystal sapphire was choosen for this SRM since Al_2O_3 is a widely used material, it has a relatively high expansion ($\alpha \sim 10 \times 10^{-6}K^{-1}$), and is stable to moderately high temperatures. A single crystal was used since the question of preferred orientation, that may arise with polycrystalline samples, could be eliminated by measurements of the orientation. Many high temperature dilatometers are constructed from polycrystalline alumina, and since the expansion of this single crystal sapphire oriented at 59° to the c axis is a close match, it will provide a very sensitive check for thermal gradients in dilatometers that often influence expansion measurements at high temperatures. This SRM will be supplied as rods 6.4 mm in diameter and 50 mm in length or other lengths by special order.

PROCEDURE

The single crystal rods for this SRM were obtained commercially. They were grown from a melt of high purity Al_2O_3, cut into 20 cm lengths, and centerless ground to 6.4 mm diameter. Since Al_2O_3 is trigonal, the expansion in a direction inclined at an angle ω from the c axis is given (2) by

$$\frac{\Delta L}{L}(\omega) = \frac{\Delta L}{L}(c) \cos^2 \omega + \frac{\Delta L}{L}(a) \sin^2 \omega \qquad (1)$$

where $\Delta L/L(c)$ is the expansion along the c-axis (the three fold inversion axis) and $\Delta L/L(a)$ the expansion perpendicular to the c-axis. These expansion measurements were made parallel to the rod axis so that ω is the inclination of the rod axis to the c-axis of the crystal. The angle ω was measured on each rod using back-reflection Laue x-ray photographs. An average ω of 59° was obtained for the rods with a standard deviation of 0.6° and a spread of from 58° to 60°. A number of rods were also checked for variations in ω along their length using the same technique, but no variations were found. Three samples were selected for the expansion measurements with orientations of 58°, 59°, and 60° in order to check for differences in expansion due to misorientation.

Expansion measurements from 293 to 1000 K were made with a Fizeau interferometer technique (3). Three 1-cm samples were cut from each rod and ground to fit an alumina sample holder. The ends of the samples were ground conical in shape with flat parallel ends having an area of about $3mm^2$ which were optically polished to provide a stable three point spacer for the fused silica optical flats. A recess hole was drilled into the sample holder for a thermocouple. The thermocouple is a Pt vs. Pt-10Rh calibrated to IPTS-68 by the Temperature Physics Section of the NBS. The calibration of this thermocouple is checked periodically to 800 K using a calibrated platinum resistance thermometer. Interference fringes were formed between the optical flats using the green spectral line of a mercury lamp and fringe motion was measured with a filar-micrometer eyepiece. The interferometer was heated in a low pressure helium atmosphere furnace and the expansion of the sample measured between equilibrium temperatures. Gage blocks were used to measure the sample length at ambient temperature. With this measurement system the expansion per unit length is believed to be accurate to 5 μm/m at 1000 K.

Measurements in the range from 1000 to 2000 K were made with a twin tele-microscope apparatus described by Rothrock and Kirby (4). These microscopes have a 50x magnification and are rigidly attached to an invar bar to minimize the influence of room temperature changes. Filar- micrometer eyepieces are used to compare the length of the sample during testing to that of a fused-silica scale at room temperature. Knife edges, ground into the sample perpendicular to the rod axis, are used for measurement of the length which is nominally 10 cm. The standard deviation of repeated settings on a knife edge is less than 1 μm. This gives an expected precision of 20μm/m in $\Delta L/L$ measurements. The sample was heated in vacuum using a tungsten-mesh furnace similar in design to that used by Burns and Hurst (5) for a study of thermocouple behavior at elevated temperatures, but with additional windows for observing the sample. A unique feature of the furnace is a radiation shield between the sample and tungsten

mesh heating element which reduced the temperature gradient in the
furnace. A probe thermocouple calibrated to IPTS-68 was used to
calibrate the measurement thermocouple and to determine the temperature gradient at one hundred degree intervals in the sample chamber.
The length of the sample was measured at equilibrium temperatures and
corrections were made for the temperature gradient. With the accuracy
of length and temperature measurements, it is believed that the
expansion per unit length is accurate to 40 μm/m at 2000 K.

RESULTS

Expansion measurements to 1000 K gave agreement in the expansion
per unit length between the three rods to within 4 μm/m. The expected
difference for an angular misorientation of one degree is about
10×10^{-6}. The standard deviation of the fitted data from the individual samples is less than 5×10^{-6}. Thus from the expansion measurements we could not infer any difference in orientation of the rods
tested. If a difference of $1/2°$ were extrapolated to 2000 K, the
expected difference in expansion would be about 10×10^{-6} which is
below the expected limits of detection of the high temperature
apparatus. As a result, it is assumed that there is no difference
in the orientation of the rod and, therefore, only one sample was
measured in the temperature range from 1000 to 2000 K. The expansion data were pooled and fitted using a procedure given by Wachtman
et al. (6) using the equation

$$\frac{L_T - L_{293}}{L_{293}} = \frac{1}{3a} \; \frac{E\left(\frac{\Theta}{T}\right)}{Q_o - kE\left(\frac{\Theta}{T}\right)} + \frac{1-a}{a} \tag{2}$$

where

$$E\left(\frac{\Theta}{T}\right) = \frac{3nR\theta}{4} \; [(e^{\theta/T} - 1)^{-1} + (e^{\theta/2T} - 1)^{-1}].$$

R is the gas constant ($8.3143 \ JK^{-1} \ mole^{-1}$), n is the number of atoms
per molecule, and a = L_{293}/L_o. The values θ = 927 K, $Q_o = 5 \times 10^6$
J/mole, k = 3.6 and a = 1.000674 were obtained by successive least
squares approximations, as described by Wachtman, et al., in order to
minimize the standard deviation of the fit. Using 113 data points,
the residual standard deviation of the fit for the parameters listed
above is 16×10^{-6}. As a further check of the homogeneity of the
three samples, the average deviation of the data for each sample from
the fitted curve was calculated resulting in values of 1.0×10^{-6},
1.8×10^{-6}, and -1.8×10^{-6}. The small magnitude of these differences
confirm that these samples are homogeneous and the lot of material is
of a suitable quality for a SRM. Since the two experimental methods
used have different sensitivities, separate residual standard deviations were calculated. In the temperature range up to 1000 K the

residual standard deviations is 5×10^{-6} and for the data above 1000 K the residual standard deviation is 25×10^{-6}. All of the data are within approximately two standard deviations of the predicted value in their respective temperature intervals. A plot of the residuals from the final curve for each sample is shown in Figure 1.

Since Equation 2 is just an approximation, as discussed by Wachtman, et al., it was also decided to determine if the high and low temperature values might adversely bias each other. In Table 1 are listed values calculated from the parameters determined by fitting the data from each sample, pooled data to 1000 K, high temperature data, and all the data. While the parameters from the different equations were generally different, the predicted expansion values from the separate ranges effect each other only to the extent of a few parts per million. Considering the standard deviation in the two ranges of 5×10^{-6} and 25×10^{-6} it is felt that this is not a significant effect.

From Equation 2, the coefficient of thermal expansion is

$$\alpha = \frac{1}{L_{293}} \frac{dL}{dT} = \frac{nR\theta^2}{32aT^2} \frac{Q_o}{(Q_o - kE)^2} \left[4 \operatorname{csch}^2 \frac{\theta}{2T} + \operatorname{csch}^2 \frac{\theta}{4T} \right] \quad (3)$$

Values calculated from this equation are also listed in Table 1.

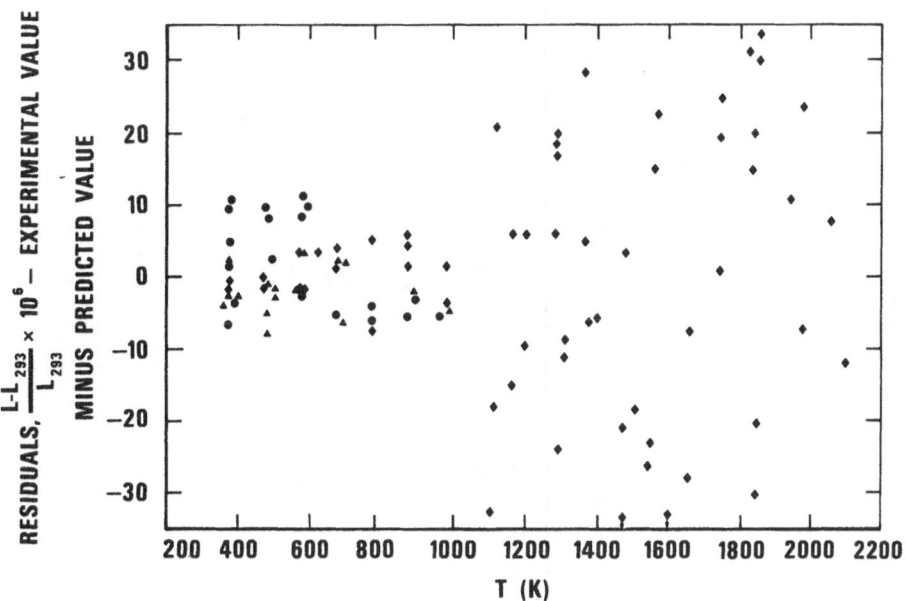

Fig. 1. Residuals from the least-square analysis of the expansion data. ◆Sample from rod #15, ●sample from rod #17, ▲sample from rod #29.

TABLE 1

CALCULATED THERMAL EXPANSION OF SINGLE CRYSTAL SAPPHIRE

$$\frac{L-L_{293}}{L_{293}} \times 10^6$$

T(K)	Interferometer Measurements				Tele-Microscope Measurements	Pooled	$\alpha = \frac{1}{L_{293}}\frac{dL}{dT}$
	Rod #15	Rod #17	Rod #29	Pooled			$\times 10^{-6}$
293	0	0	0	0		0	5.38 $\times 10^{-6}$
300	37	37	37	37		38	5.48
400	644	652	643	646		647	6.61
500	1343	1350	1342	1345		1346	7.32
600	2104	2105	2101	2103		2104	7.82
700	2907	2902	2904	2904		2906	8.20
800	3744	3738	3740	3740		3743	8.52
900	4609	4605	4603	4605		4608	8.79
1000	5498	5494	5495	5497		5500	9.05
1100					6414	6417	9.29
1200					7355	7357	9.52
1300					8318	8320	9.74
1400					9304	9306	9.97
1500					10300	10310	10.20
1600					11345	11345	10.42
1700					12400	12400	10.66
1800					13480	13480	10.89
1900					14580	14580	11.14
2000					15710	15700	11.38

The differences between the coefficients calculated from the experimental data and the values calculated from Equation 2 are shown in Figure 2. In the temperature range to 1000 K the calculated standard deviation is 0.04×10^{-6}/K and from 1000 to 2000 K the standard deviation is 0.18×10^{-6}/K. Examination of Figures 1 and 2 show that there are no systematic deviations of the data from these equations and that these equations provide a good representation of the data. In contrast to the values of $\theta = 927$ and $k = 3.6$ found in this study, Wachtman et al., determined θ to be of the order of 1100 K and k<1 to best fit their Al_2O_3 data on the expansion parallel and perpendicular to the principal directions. Their value is more characteristic of a Debye θ found for fitting the low temperature heat capacity data of Furukawa et al. (7) while $\theta = 927$ is more characteristic of the average value of 950 K (6) found for fitting the high temperature heat capacity data. The value of $k = 3.7$ found here is also characteristic of materials studied by Grüneisen (8) who found k generally greater than 2. The value of k is related to the exponential factors in the force constants between atoms.

A comparison is shown on Table 2 of the expansion values found in this study to those values found in the literature. When the values were not at the temperatures listed here, they were fitted with a polynomial and then values calculated at the temperatures listed in Table 2. All of the temperatures have been corrected to IPTS-68

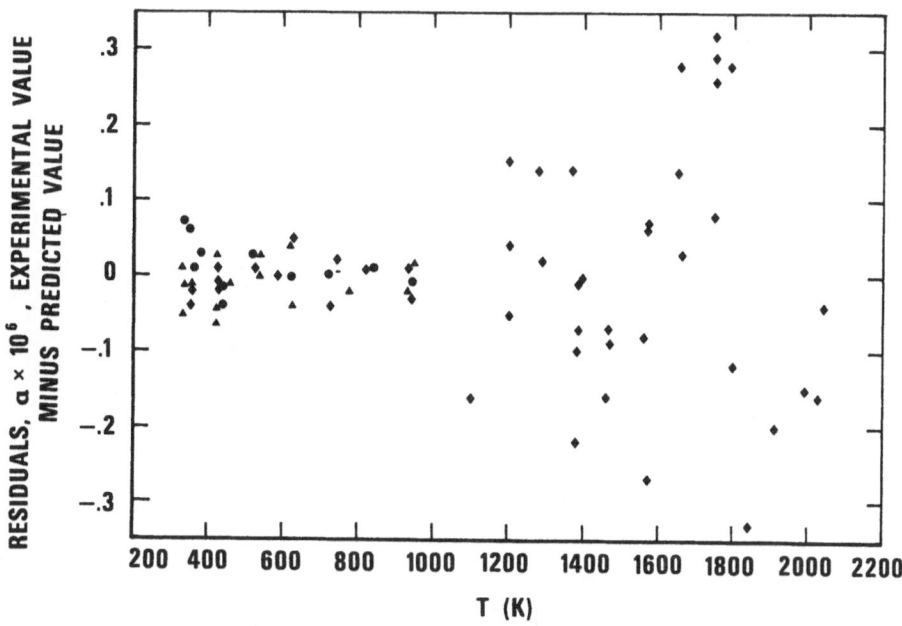

Fig. 2. Calculated residuals for the expansivity ($\alpha = 1/L_{293}$ dL/dT) using predicted values from Eq. 3. ◆ sample from rod #15, ● sample from rod #17, ▲ sample from rod #29.

TABLE 2

THERMAL EXPANSION OF SINGLE CRYSTAL SAPPHIRE

$$\frac{L - L_{293}}{L_{293}} \times 10^6 \text{ for } \omega = 59°$$

T(K)	Present Study	Schauer [11]	Wachtman et al. [16]	Amatuni and Shevchenko [12]	Yates et al [13]	Agard [14]	Ebertand Tingwaldt [15]	Mauer and Boltz [16]	Campbell and Grain [17]	$\Delta L/L(\omega=59°) - \Delta L/L$(polycrystalline)
293	0	0	0	0	0	0	0	0	0	0
300	38	37	35	35	37	39				− 1
400	647	644	628	635	630	649		625	661	− 6
500	1346	1346	1332	1342	1342	1336				− 11
600	2104	2105	2105	2108	2090	2090	2106	2153	2162	− 17
700	2906	2904	2924	2918	2906	2899				− 23
800	3743	3739	3774	3768	3752	3753	3720	3807	3915	− 29
900	4608	4606	4646	4653		4641				− 35
1000	5500	5502	5530	5545		5553	5511	5420	5820	− 40
1100	6417	6426	6427			6475				− 46
1200	7360	7370	7330			7396	7420	7020	7820	− 53
1300	8320	8340	8230			8308				− 59
1400	9310	9320	9140			9340	9410	8700	10030	− 65
1500	10310		10050			10330				− 71
1600	11340		10960			11340	11400	11190		− 77
1700	12400		11870			12350				− 83
1800	13480		12780				13420			− 90
1900	14580		13700							− 96
2000	15700		14620				15530			−101

(Wachtman et al. values for the upper temperatures are marked "Extrapolated".)

using the recommended values given by Preston-Thomas (9) for $t_{68}-t_{48}$
and Corruccini (10) for $t_{48}-t_{27}$. If values were given for the
expansion of the parallel and perpendicular directions. Equation 1
was used to calculate the expansion for $\omega = 59°$. Corrections to
obtain the expansion relative to the length at 293 K were obtained
from the data of Wachtman et al. Listed in Table 2 are values used
to calculate the expansion for $\omega = 59°$ from data on polycrystalline
samples. These values were also calculated from the data of Wachtman
et al. including their extrapolated data in the range from 1100 to
2000 K taking

$$\Delta L/L \text{ (polycryst)} = 1/3 \left[\Delta L/L(c) + 2\Delta L/L(a)\right] \text{ and } \Delta L/L(\omega = 59°)$$

from Equation 1.

From 293 to 1000 K the correction factor agreed to within
1×10^{-6} when calculated from the data of either Wachtman et al. or
Amatuni and Shevchenko. Above 1000 K this correction factor is
probably accurate to 10% since it only depends upon the relative
expansion in the c and a directions which still allows a meaningful
comparison between the different data.

Schauer (11) measured the expansion of polycrystalline Al_2O_3
from 100 to 1470 K using sapphire interferometer plates. In order
to get L_T-L_{293}/L_{293} from the reported thermodynamic coefficients
($\alpha = 1/L \ dL/dT$), the coefficients were fitted with a fifth degree
polynomial and the expansion calculated from $L_T-L_{293}/L_{293} =$
$[\exp \int_{293}^{T} \alpha \ dT] -1$. The excellent agreement with the values found
in this study help confirm the absolute expansion of this material.
In addition, the corrections for the expansion difference between the
polycrystalline and $\omega = 59°$ orientation are confirmed.

The agreement with the other interferometric data on single
crystal measurements of the mutually perpendicular axes of Wachtman
et al., Amatuni and Shevchenko (12) and Yates et al. (13) is generally
within 1% up to 1000 K. The extrapolation of Wachtman et al. from
1100 to 2000 K results in values lower than found in this study by
approximately 7% at 2000 K.

The AGARD results (14) are from a cooperative study involving a
number of different laboratories using fused quartz dilatometry to
1200 K and alumina dilatometers at higher temperatures. Above 1200 K
a group average was calculated from the data presented for the six or
seven laboratories participating in the high temperature program.
While the spread in the data was large (as high as 8%) the conclusion
given that the group average was close to the true expansion is confirmed by the good agreement with the other values given in Table 2.

Ebert and Tingwaldt's data (15) using a twin-telescope technique
were taken on a polycrystalline material. The agreement with the

value from the present study is about 1% at 2000 K, just slightly greater than the combined uncertainties.

Also listed in Table 2 are the x-ray diffraction studies of Mauer and Bolz (16) and Campbell and Grain (17). Temperature corrections to IPTS-68 were not made on these data. These data deviate above and below the other values listed in Table 2 so that the variations cannot be attributed to differences between bulk and lattice measurements. The present study will help to establish Al_2O_3 as an expansion standard for x-ray equipment even though measurements were made on only one orientation. If the extrapolated data of Wachtman et al. and the various x-ray studies are used to calculate an average relative expansion between the c and a axis directions and the expansion data for the 59° orientation from this study is used to establish the magnitude, the lattice expansion can probably be determined with a smaller uncertainty than if the data from only one of the x-ray studies is used. Because of the uncertainties in this procedure, there is still a need for anisotropic expansion data on single crystal Al_2O_3 above 1000 K.

REFERENCES

1. These Standard Reference Materials are available from the Office of Standard Reference Materials, National Bureau of Standards, Washington, D. C. 20234.

2. J. F. Nye, Physical Properties of Crystals - Their Representation by Tensors and Matrices. Oxford University Press, New York, 1957.

3. T. A. Hahn, J. Appl. Phys. 41, 5096 (1970).

4. B. D. Rothrock and R. K. Kirby, J. Research Natl. Bur. Standards 71C, 85 (1967).

5. G. W. Burns and W. S. Hurst, NASA Report CR-72884, 1972.

6. J. B. Wachtman, T. G. Senderi, and G. W. Cleek, J. Am. Ceram. Soc. 45, 319 (1962).

7. G. T. Furukawa, T. B. Douglas, R. E. McCoskey, and D. C. Ginnings, J. Research Natl. Bur. Standards 57, 67 (1956).

8. E. Grüneisen, NASA Publ. No. RE 2-18-59W (translation of Handbuch der Physik, Vol. 10, 1-52 (1926)).

9. H. Preston-Thomas, Metrologia 12, 7 (1976).

10. R. J. Corruccini, J. Research Natl. Bur. Standards 43, 133 (1949).

11   A. Schauer, Can. J. Phy. 43, 523 (1965).

12.   A. N. Amatuni and E. B. Shevchenko, Measurement Techniques 15, 10 (1966). (Translation of Izmeritel Naya Tekhnika 10, 17 (1966)).

13.   B. Yates, R. F. Cooper, and A. F. Pojur, J. Phys. C; Solid State Phys. 5, 1046 (1972).

14.   E. Fitzer, AGARD Advisory Report No. 38, NATO, 1972.

15.   Von H. Ebert and C. Tingwaldt, Physik, Z. 37, 471.

16.   F. A. Mauer and L. H. Bolz. WADC Tech, Rept. 55-473, 1958, ASTIA Document No. 155555.

17.   W. J. Campbell and C. Grain, Advances in X-ray Analysis, Wm. M. Mueller editor, Vol. 5, 244, 1961, Plenum Press, New York.

THERMAL DEFORMATIONS AND RESIDUAL STRESSES IN FIBER COMPOSITES

I.M. Daniel

IIT Research Institute

10 W. 35th Street, Chicago, Illinois   60616

ABSTRACT

The basic composite lamina is highly anisotropic thermally. Thermal expansion is lowest in the fiber direction and highest in the matrix-dominated transverse direction. The low coefficient of thermal expansion (α) and high modulus in the fiber direction, especially in graphite composites, allow fabrication of angle-ply laminates with near-zero thermal expansion. This is of great importance in structures requiring exceptional dimensional stability. The coefficient α for a single ply can be calculated knowing the thermal and mechanical properties of the constituents and their geometric distribution. Thermal expansion in angle-ply laminates can also be predicted reasonably well by means of lamination theory. An important result of thermal anisotropy is the introduction of lamination residual stresses in angle-ply laminates during curing. They can reach values comparable to the transverse strength of the ply and thus induce cracking of that ply within the laminate. They are equilibrated with interlaminar shear stresses transmitted from adjacent plies and can thus cause delamination. Residual stresses have been investigated analytically and experimentally. It was found that the significant strains recorded during the cooling stage of curing correspond to thermal expansion of the laminate. Residual or restraint strains are computed from measured restrained and unrestrained thermal expansions. Residual stresses are computed using appropriate orthotropic constitutive relations. Results

have been obtained for a variety of materials including
boron, graphite, Kevlar, S-glass and hybrids with epoxy
or polyimide matrices, for a variety of lamination angles.
It was found that residual stresses do not relax appreciably with time. Results show that, for graphite and
Kevlar laminates, residual stresses at room temperature
are high enough to have caused damage in the transverse
to the fiber direction.

INTRODUCTION

The high strength, high stiffness, low density and
good fatigue endurance of most composites make them
desirable for many critical structural applications. In
designing with composites most of the emphasis to date
has been placed on mechanical properties and less on
fundamental physical properties. One of the most
important physical properties which has a bearing on the
strength and function of composite structures is the coefficient of thermal expansion, α.

The basic composite lamina or ply, consisting of an
array of parallel fibers in a matrix, is highly anisotropic thermally, as the response in the fiber direction
is dominated by the stiff low-expansion fibers and the
response in the transverse direction is governed by the
low stiffness high-expansion matrix. This thermal anisotropy is most pronounced in Kevlar and graphite composites
where the coefficient of thermal expansion is usually
negative in the fiber direction. The low α and high
modulus in the fiber direction, especially in graphite/
epoxy composites, allow the fabrication of angle-ply
laminates with near-zero thermal expansion. This is of
great importance in structures requiring exceptional
dimensional stability, such as optical mirrors and
structures, ultrahigh frequency antenna reflectors,
radiation shields, stabilizing platforms, focusing devices,
metering rods in optical systems, etc. Another important
result of the anisotropic thermal expansions of composite
plies is the introduction of lamination residual or
thermal stresses in angle-ply laminates during curing.
In the design and evaluation of composite structures one
must take these residual stresses into account and superimpose them unto those produced by subsequent external
loading and environmental fluctuations.

This paper deals with measurement methods and results
on thermal expansion characteristics and the determination
of lamination residual stresses in fiber composites.

THERMAL DEFORMATIONS

Theory

The coefficient of thermal expansion α for a single ply or a unidirectional laminate can be calculated by knowing the coefficients of the fiber and matrix, their geometric arrangement and mechanical properties.[1,2] Schapery's[1] energy theory yields the following expressions for a unidirectional fibrous composite:

$$\alpha_{11} = \frac{\alpha_f E_f V_f + \alpha_m E_m V_m}{E_f V_f + E_m V_m}$$

$$\alpha_{22} = \alpha_f V_f (1+\nu_f) + \alpha_m V_m (1+\nu_m) - \alpha_{11} (\nu_f V_f + \nu_m V_m)$$

where α_{11}, α_{22} = coefficients of thermal expansion for composite lamina in the fiber and transverse to the fiber directions, respectively.

α_f, α_m = coefficients of thermal expansion for fiber and matrix, respectively.

E_f, E_m = fiber and matrix moduli, respectively.

V_f, V_m = fiber and matrix volume fractions, respectively.

ν_f, ν_m = fiber and matrix Poisson's ratios, respectively.

For realistic graphite/epoxy composites with very low α_f and high E_f the expression for the transverse coefficient of expansion can be approximated by

$$\alpha_{22} \cong \alpha_f V_f + \alpha_m V_m (1+\nu_m)$$

The expansion coefficients along an arbitrary coordinate system x-y rotated by an angle θ with respect to the fiber direction can be obtained by a linear coordinate transformation as

$$\alpha_x = \alpha_{11} \cos^2\theta + \alpha_{22} \sin^2\theta$$

$$\alpha_y = \alpha_{11} \sin^2\theta + \alpha_{22} \cos^2\theta$$

From the coefficients of a single lamina the effective coefficients for angle-ply laminates can be evaluated
by means of lamination theory.[3-5] Computer programs exist
for such calculations. Most of these theories are linear
elastic and do not take into consideration the viscoelastic response of the matrix at elevated temperatures.
Coefficients of thermal expansion for a general
$[0/\pm45/90]_c$ boron/epoxy laminate at room temperature and
at the curing temperature of 450 degK (350°F) are shown
in Fig. 1 as a function of ply orientation and ply
composition of the laminate.[4] Thus, for a ply configuration of $[0_2/\pm45]_c$ the coefficients of thermal expansion
at room temperature would be

$$\alpha_{11} = 4.0 \times 10^{-6} K^{-1} (2.2\mu\epsilon/°F)$$

$$\alpha_{22} = 10.8 \times 10^{-6} K^{-1} (6.0\mu\epsilon/°F)$$

These coefficients can be compared with the unrestrained
single-ply coefficients for the 0-deg. ply

$$\alpha_{11} = 4.5 \times 10^{-6} K^{-1} (2.5\mu\epsilon/°F)$$

$$\alpha_{22} = 23.4 \times 10^{-6} K^{-1} (13\mu\epsilon/°F)$$

The differences between the coefficients of the angle-ply
laminate and those of the single ply give a measure of the
residual strains present in that ply within the laminate.

Experimental Techniques

Thermal expansion in unidirectional and angle-ply
laminates has been measured experimentally for various
materials and over wide temperature ranges. Freeman and
Campbell[6] used a Leitz dilatometer and strain gages to
measure thermal expansion in graphite fiber composites
over a temperature range of 78 to 561°K (-320 to 550°F).
Freund[7] used a vacuum interferometric dilatometer to
measure thermal expansion in high-modulus graphite/epoxy
laminates for applications to mirror mounts requiring
high dimensional stability. An accuracy of $\pm1 \times 10^{-8}/°K$
was mentioned. Thermal expansion has also been measured
by Wang et al.[8] and the author[9,10,11] for a variety of
composite materials using strain gages. Commercial
strain gages have been found suitable for measuring low
thermal strains in composites. To properly interpret the
strain gage output ϵ_a (apparent strain), it is necessary
to separate this output into the component ϵ_t due to the
deformation of the specimen (thermal strain) and the
component ϵ_g due to changes in resistivity and gage

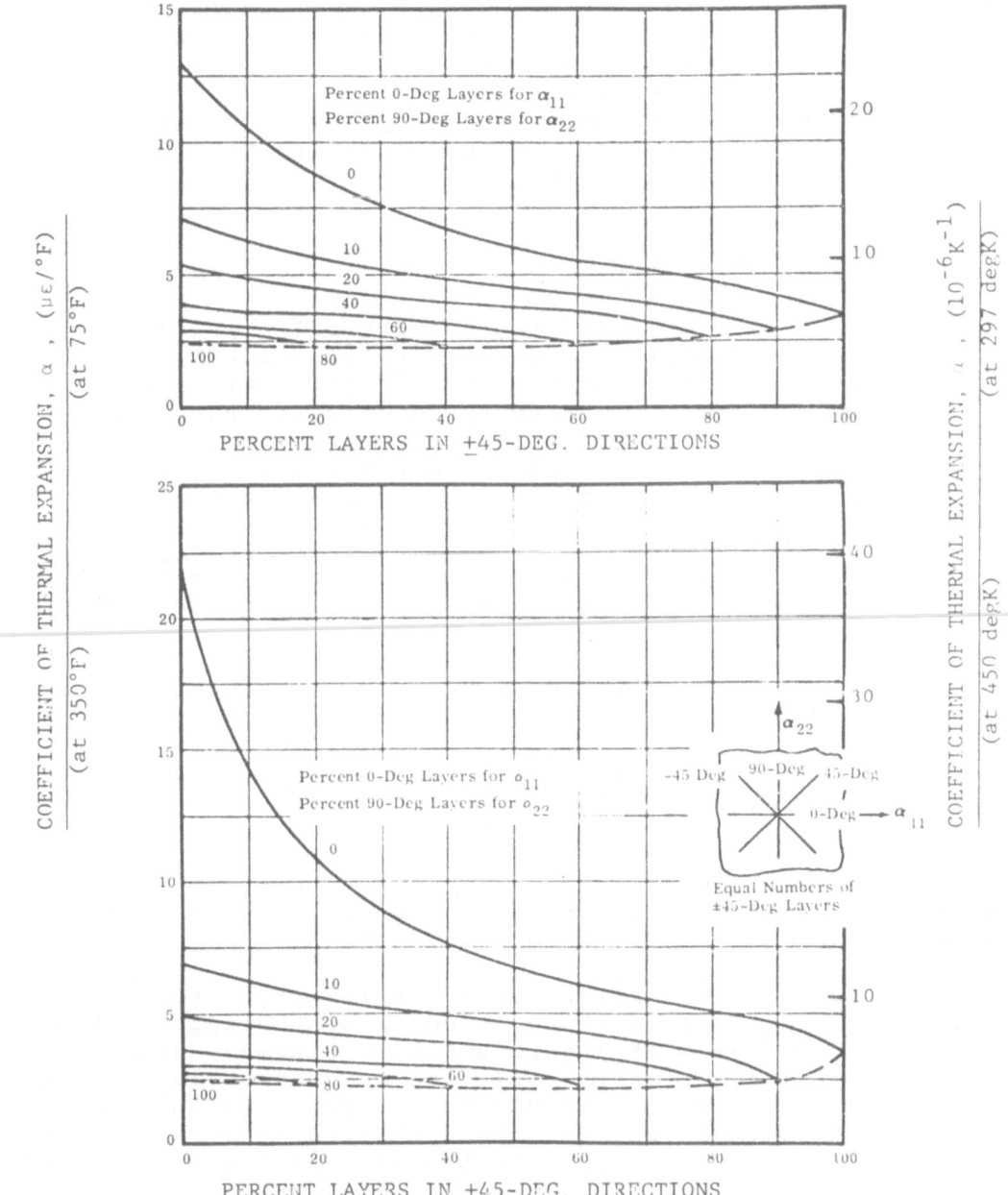

Fig. 1   Coefficients of Thermal Expansion for
          $[0/90/\pm45]_c$ Boron/Epoxy Laminates at
          Room Temperature and 450° degK
          (350°F) (Ref. 4)

factor of the gage with temperature (thermal output).
To determine ε_g a reference specimen of known thermal
expansion is instrumented with the same type of gage as
the specimens to be tested. The true thermal strain ε_t
is then obtained by subtracting algebraically from the
recorded apparent strains ε_a the output of the gage on
the reference material ε_r and adding the known expansion
of the reference material ε_{tr}

$$\varepsilon_t = \varepsilon_a - \varepsilon_r + \varepsilon_{tr}$$

Fused quartz with a coefficient of thermal expansion of
$0.7 \times 10^{-6}K^{-1}$ ($0.4\mu\varepsilon/°F$) and titanium silicate with a
coefficient of $0.03 \times 10^{-6}K^{-1}$ ($0.017\mu\varepsilon/°F$) have been used
as reference materials. Many types of gages have been
used. In the case of graphite/epoxy composites and
especially when gages are embedded between plies it is
necessary to use fully encapsulated gages with insulated
leads. Micro-Measurements gages of the WK-00 series were
found most suitable as they have a very low purely thermal
output. The apparent strain recorded with a typical gage
of this series on a titanium silicate specimen is shown
in Fig. 2 as a function of temperature.

    Strain gages record local deformations averaged over
their gage length, therefore they tend to reflect local
material irregularities, inhomogeneities and flaws in a
realistic composite. For this reason some variations may
be seen between gage readings from different gages at
different locations. These strain variations can be of
the order of ±100 $\mu\varepsilon$. Readings from surface gages may be
different from those of embedded gages because of laminate
bending. Realistic results are obtained by averaging
readings from a number of embedded gages.

    In one recent application of strain gages to measure-
ment of thermal deformations,[11] encapsulated gages (WK-00-
125TM-350, Option B-157) were embedded between the plies
during laminate assembly. The attached ribbon leads were
sandwiched between thin (0.013 mm; 0.0005 in.) polyimide
strips. A thermocouple was also embedded in each laminate.
The instrumented specimens, including a reference titanium
silicate specimen, were subjected to the curing and postcuring cycles in the autoclave. Strain gage and thermocouple readings were taken throughout. Subsequently,
the same specimens were subjected to a thermal cycle from
room temperature to 444 degK (340°F) and down to room
temperature. Strain gages and thermocouples were recorded

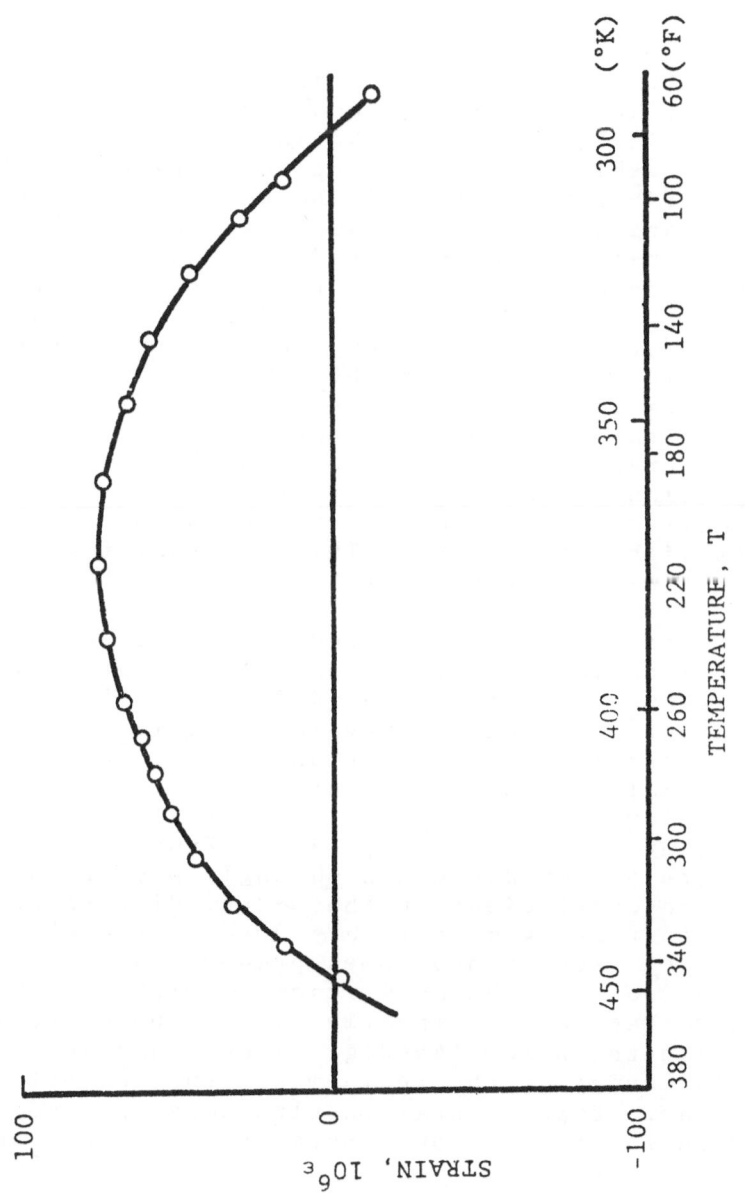

Fig. 2 Apparent Strain as a Function of Temperature of WK-00-125TM-350 Gage Bonded on Titanium Silicate

at 5.5 degK (10° F) intervals. The true thermal strains
were obtained from recorded apparent strains as discussed
above.

Thermal Strains

Thermal strains measured on eight-ply unidirectional
graphite/epoxy, Kevlar 49/epoxy and S-glass/epoxy specimens are shown in Fig. 3. Both Kevlar 49/epoxy and
graphite/epoxy exhibit negative thermal strains in the
longitudinal (fiber) direction. The Kevlar 49/epoxy
exhibits the largest positive transverse and negative
longitudinal strains. The S-glass/epoxy undergoes the
lowest thermal deformation in the transverse direction
and the highest (positive) in the longitudinal direction.
Coefficients of thermal expansion computed from such data
are tabulated in Table 1 for eight composite materials.[10,11]
Coefficients are listed at room temperature, 297 degK
(75°F), and at the elevated temperature of 450 degK
(350°F). All graphite fiber composites exhibit negative
thermal expansion in the fiber direction. The polyimide
matrix composites do not show any variation of thermal
coefficients with temperature. This is true at least up
to the postcuring temperature of 589 degK (600°F).

Freeman and Campbell[6] measured the longitudinal and
transverse dilatation of five graphite composites over a
temperature range of 78 to 561 degK (-320 to 550°F). The
moisture content was found to have a significant effect
on thermal deformation with a permanent shrinkage resulting after thermal cycling to elevated temperature. After
sufficient drying the coefficient of thermal expansion
remains constant with thermal cycling. Expansion coefficients were also measured in $\pm\theta$ angle-ply graphite
laminates. The coefficient in the 0-deg. direction decreases with increasing θ up to θ = 30-deg, thereafter
the influence of the resin becomes predominant as shown
by a rapid increase in the coefficient α with θ. Thermal
deformation varies fairly linearly with temperature for
angle-ply laminates below $\pm$45-deg. Above this angle the
increasing influence of the resin makes thermal deformations nonlinear. Experimental results were in good
agreement with theoretical ones obtained by using lamination theory.[3] Orthotropic laminates such as [0/90]$_c$,
[$\pm$45]$_c$ and [0/90/$\pm$45]$_c$ have equal coefficients along two
principal material axes. Theoretically, all three types
of laminates above are supposed to have the same coefficients along the 0-deg and 90-deg directions. However,
experimental measurements show that the [0/90]$_c$ laminate
has slightly lower thermal expansion than the other two

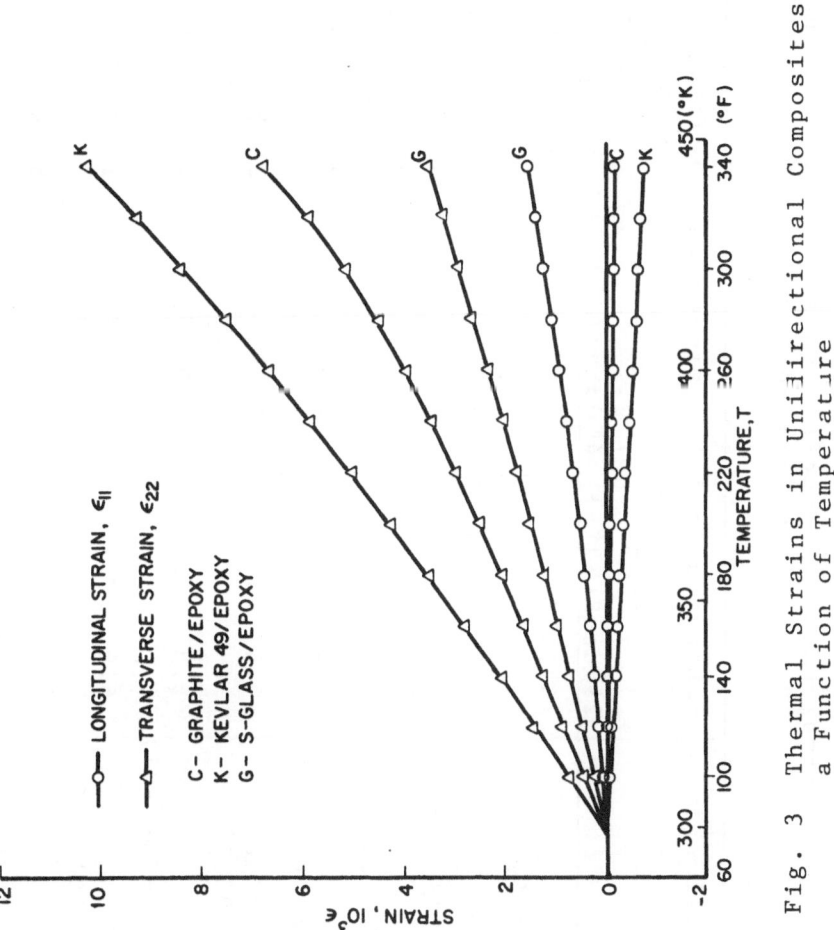

Fig. 3   Thermal Strains in Unidirectional Composites as a Function of Temperature

Table 1

Thermal Expansion Coefficients of Unidirectional Composite Materials

Material	Longitudinal Coefficient of Thermal Expansion, α_{11}, $10^{-6} \mathrm{K}^{-1}$ ($\mu\varepsilon/°\mathrm{F}$)		Transverse Coefficient of Thermal Expansion, α_{22}, $10^{-6} \mathrm{K}^{-1}$ ($\mu\varepsilon/°\mathrm{F}$)	
	297degK(75°F)	450degK(350°F)	297degK(75°F)	450degK(350°F)
Boron/Epoxy (Boron/AVCO 5505)	6.1 (3.4)	6.1 (3.4)	30.3 (16.9)	37.8 (21.0)
Boron/Polyimide (Boron/WRD 9371)	4.9 (2.7)	4.9 (2.7)	28.4 (15.8)	28.4 (15.8)
Graphite/Epoxy (Modmor I/ERLA 4289)	-1.1 (-0.6)	3.2 (1.3)	31.5 (17.5)	27.0 (15.0)
Graphite/Epoxy (Modmor I/ERLA 4617)	-1.3 (-0.7)	-1.3 (-0.7)	33.9 (18.8)	83.7 (46.5)
Graphite/Polyimide (Modmor I/WRD 9371)	-0.4 (-0.2)	-0.4 (-0.2)	25.3 (14.1)	25.3 (14.1)
S-Glass/Epoxy (Scotchply 1009-26-5901)	3.8 (2.1)	3.8 (2.1)	16.7 (9.3)	54.9 (30.5)
S-Glass/Epoxy (S-Glass/ERLA 4617)	6.6 (3.7)	14.1 (7.9)	19.7 (10.9)	26.5 (14.7)
Kevlar/Epoxy (Kevlar 49/ERLA 4617)	-4.0 (-2.2)	-5.7 (-3.2)	57.6 (32.0)	82.8 (46.0)

because of the higher percentage of fibers in the
directions of measurement. High modulus fibers produce
laminates with low thermal coefficients. These coefficients are reduced further by increasing the fiber
volume ratio. Theoretically, a quasi-isotropic laminate
of $[0/\pm60]_c$ layup with high modulus fibers and a fiber
volume ratio of 0.62 would result in zero thermal expansion. With the exception of unidirectional composites,
the thermal coefficient was found to increase nonlinearly
with temperature.

A review of thermal and other properties of
composites in the cryogenic temperature range has been
given by Kasen.[12] Results by Toth et al.[13] for glass/
epoxy in the 5 to 295 degK (-450 to 71°F) range show that
the coefficient of thermal expansion decreases with
temperature approaching zero near absolute zero. Longitudinal thermal deformations in the cryogenic temperature
range were discussed for a variety of advanced composites.[12] Borsic/Aluminum and boron/epoxy show contraction
with decreasing temperature. Graphite/epoxy shows very
little thermal expansion (small negative coefficient) in
the fiber direction down to 77 degK (-417°F). Thereafter
there seems to be a slight reversal in contraction. PRD-
49/epoxy (Kevlar/epoxy) shows appreciable thermal
expansion with decreasing temperature with the (negative)
coefficient of thermal expansion approaching zero near
absolute zero.

RESIDUAL STRESSES

An important result of the anisotropic thermal
expansions of composite plies is the introduction of
lamination residual or thermal stresses in angle-ply
laminates during curing. These stresses have been investigated recently both analytically and experimentally.[9-11,14-16] They are a function of many parameters,
such as type of fiber and matrix, fiber volume ratio, ply
orientation, curing temperature and other variables. They
can reach values comparable to the transverse strength of
the ply and thus induce cracking of that ply within the
laminate. They are equilibrated with interlaminar shear
stresses transmitted from adjacent plies and can thus
cause delamination. In the design and evaluation of
composite structures one must take residual stresses into
account and superimpose them unto those produced by subsequent external loading and environmental fluctuations.

Residual stresses during curing have been measured

in a variety of angle-ply laminates using embedded strain
gage techniques.[9-11] Unidirectional and angle-ply specimens were instrumented with surface and embedded gages
and thermocouples and the output recorded during curing
and postcuring. The unidirectional specimen was used as
a reference to determine the unrestrained stress-free
thermal expansions of an individual ply.

It was found that apparent strains recorded during
the heating stage of the curing cycle are not significant
as they correspond to the fluid state of the matrix resin.
Residual stress buildup occurs only upon solidification
of the matrix at the peak curing temperature and during
subsequent cooldown. Strains measured during the cooldown
stage of curing as well as those measured during postcuring
correspond to the thermal expansion of the laminate.

Thermal strains measured in a unidirectional graphite/
epoxy laminare are shown in Fig. 4a with room temperature
as the reference level. Thermal strains measured in a
$[0_2/\pm 45]_s$ graphite/epoxy angle-ply laminate during the
cooling stage of postcuring are shown in Fig. 4b. The
residual stresses induced in each ply correspond to the
so-called restraint strains, i.e., the difference between
the unrestrained thermal expansion of that ply (obtained
from the unidirectional specimen) and the restrained
expansion of the ply within the laminate (obtained from
the angle-ply specimen). Restraint or residual strains
in the 0- and 45-deg. plies of the $[0_2/\pm 45]_s$ graphite/
epoxy laminate are plotted in Figs. 5a and 5b as a function of temperature with room temperature taken as the
reference level. The stress-free level can be shifted
to 444 degK (340°F), the temperature at which the matrix
solidifies. Other investigators have claimed that the
stress-free temperature level might be somewhat lower
than the peak curing temperature as indicated by comparing experimental and theoretical results.[17] In the
case in question the maximum residual strain at room
temperature is 6.43×10^{-3} in the ± 45-deg. plies in the
transverse to the fiber direction. The corresponding
maximum residual strain in the 0-deg. plies is 5.95×10^{-3}.

Residual stresses in any given ply at any given
temperature can be computed from the residual strains
using the appropriate orthotropic constitutive relations.
Assuming linear elastic behavior, these relations take
the form

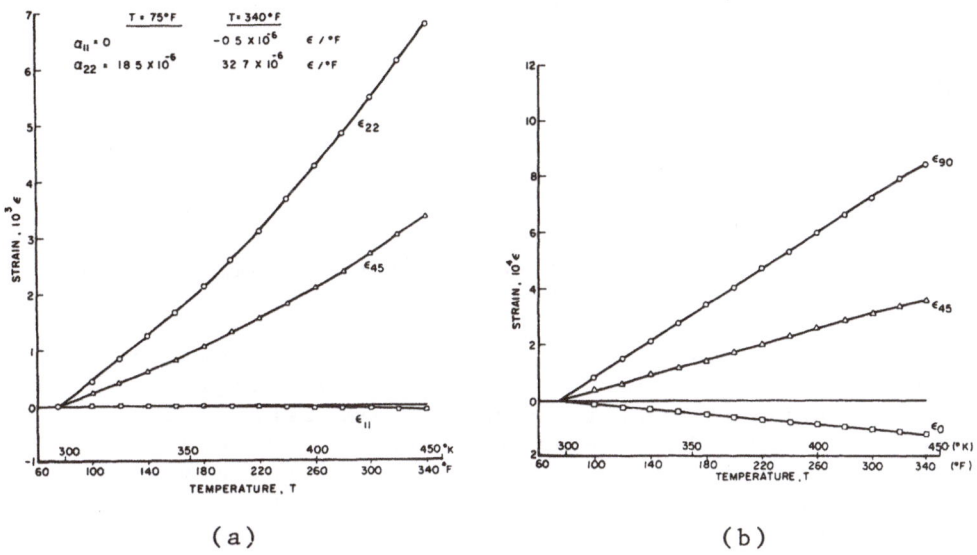

(a)   (b)

Fig. 4   Thermal Strains in Graphite/Epoxy Specimens, (a) $[0_8]$ Unidirectional Specimen, (b) $[0_2/\pm45]_s$ Specimen

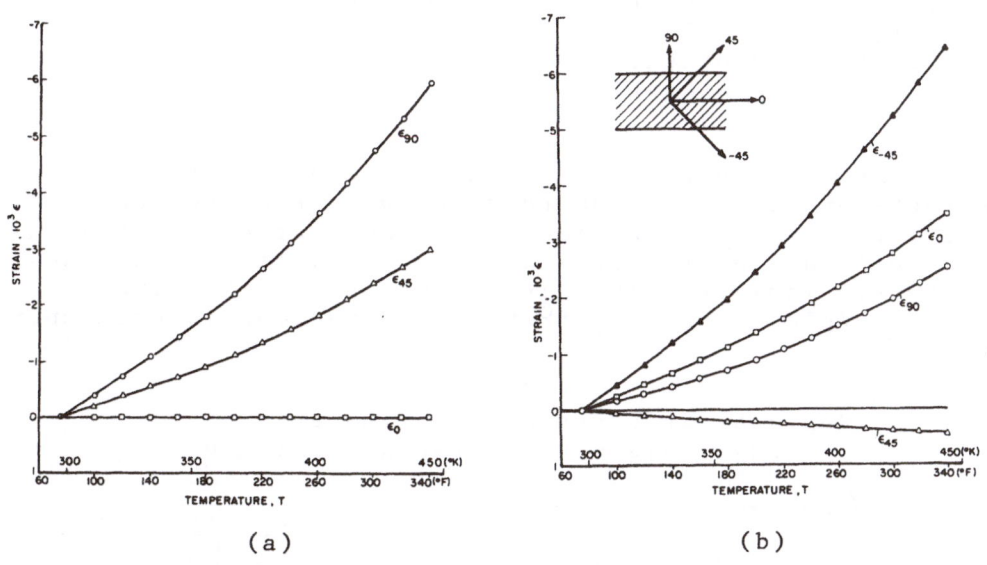

(a)   (b)

Fig. 5   Residual Strains in $[0_2/\pm45]_s$ Graphite/Epoxy Specimen.   (a) 0-Degree Plies, (b) 45-Degree Plies

$$\sigma_{11} = \frac{E_{11}}{1-\nu_{12}\nu_{21}} \left[\varepsilon_{11} + \nu_{21}\,\varepsilon_{22}\right]$$

$$\sigma_{22} = \frac{E_{22}}{1-\nu_{12}\nu_{21}} \left[\varepsilon_{22} + \nu_{12}\,\varepsilon_{11}\right]$$

$$\sigma_{12} = 2\,G_{12}\,\varepsilon_{12}$$

where superscripts 1 and 2 refer to the fiber and the transverse to the fiber directions.

Residual stresses at room temperature computed for the 0-deg. and ±45-deg. plies of the $[0_2/\pm45]_s$ graphite/ epoxy laminate at room temperature are tabulated in Table 2. The transverse to the fibers stress in the ±45-deg. plies seems to exceed somewhat the measured transverse tensile strength of the unidirectional material which is 42 MPa (6.1 ksi). This means that these plies are probably damaged in their transverse direction upon completion of curing.

Comparable results for glass/epoxy and boron/epoxy show that residual tensile stresses exhaust a significant portion of the transverse tensile strength of the ply. In the case of Kevlar 49/epoxy computed residual stresses, assuming linear elastic behavior, far exceed the trans- verse strength of the ply. It has been shown also that the amount of relaxation of residual stresses is fairly small.[18]

Further experimental work has been reported on the effects of laminate construction, ply stacking sequence and interply hybridization on residual stresses.[19,20] The influence of residual stresses on failure patterns of hybrid laminates is illustrated in Fig. 6. The graphite/glass/epoxy $[\pm45^C/0^G/0^C]_s$ specimen (where super-

Table 2

Residual Stresses at Room Temperature in
$[0_2/\pm45]_s$ Graphite/Epoxy Laminate

Ply (deg)	Stress, MPa (ksi)		
	σ_{11}	σ_{22}	σ_{12}
0	23 (3.3)	42 (6.1)	0
±45	-52 (-7.5)	45 (6.5)	6 (0.9)

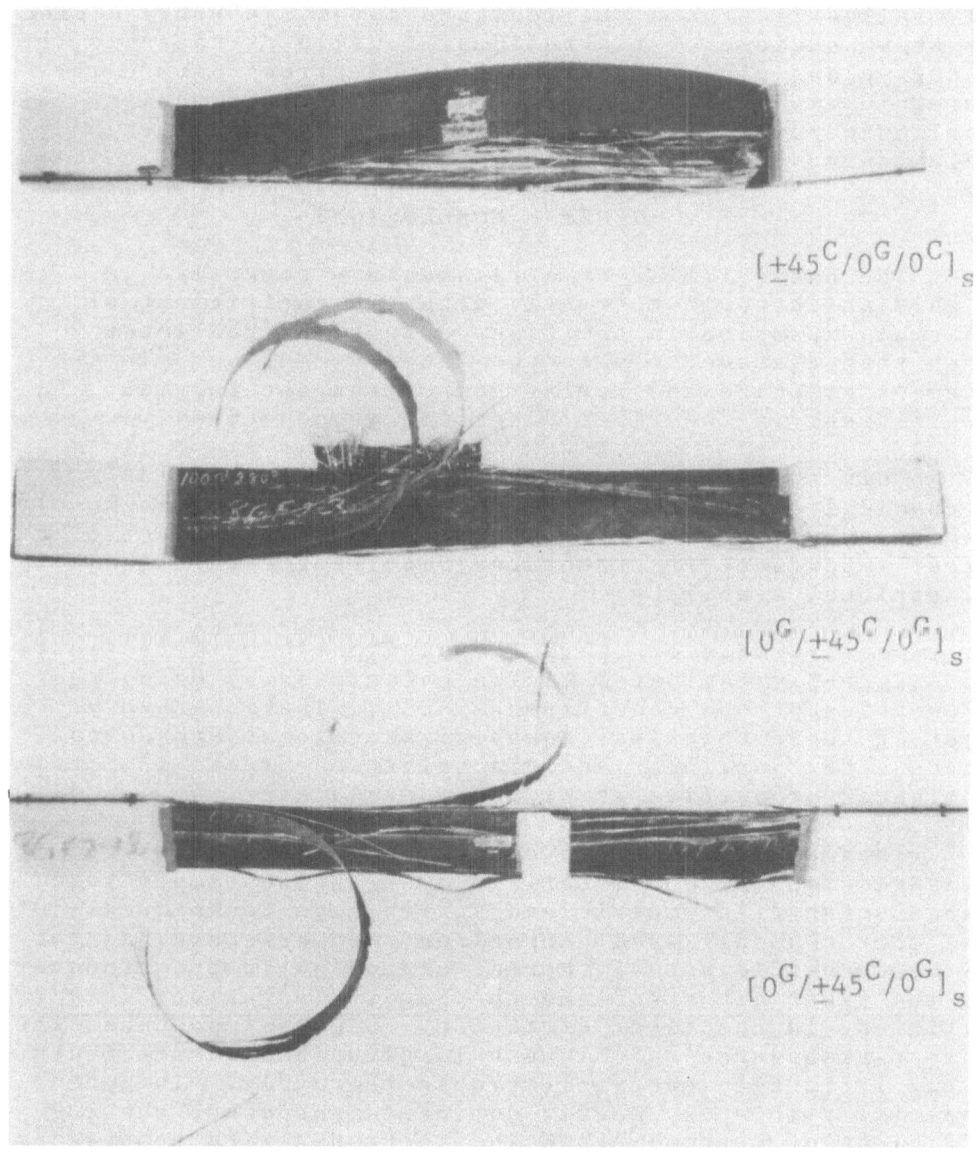

Fig. 6   Characteristic Failure Patterns of Three
          Graphite/S-Glass/High Modulus Epoxy
          Specimens Under Uniaxial Tensile Loading

scripts C and G denote graphite and glass, respectively)
failed in a "brooming" fashion after the 0-deg. graphite
ply failed first and the isolated layer ($\pm 45^C/0^G$) curled
as shown because of the residual tensile stresses in the
0-deg. glass ply. The same residual stress caused the
outer ($0^G/\pm 45^C$) layer in the $[0^G/\pm 45^C/0^G]_s$ specimens to
curl outwards after the central 0-deg. glass plies had
delaminated.

GENERAL CONCLUSIONS

The basic lamina of a filamentary composite is
highly anisotropic thermally with the coefficient of
thermal expansion in the fiber direction much lower
than that transversely to the fiber direction. In the
case of graphite and Kevlar composites the thermal
coefficient in the fiber direction is negative.

This thermal anisotropy and the corresponding
mechanical anisotropy allow the fabrication of angle-ply
laminates with near-zero thermal expansion, a fact of
great importance in structures requiring exceptional
dimensional stability.

Coefficients of thermal expansion in unidirectional
and angle-ply laminates can be calculated by using the
properties of the constituents and lamination theory.
Most of these theories, however, are linear and do not
account for nonlinear and viscoelastic effects in matrix
dominated properties at high temperatures.

Thermal expansion in composite laminates has been
measured for various materials over a wide temperature
range using dilatometric and strain gage techniques. Of
the most commonly used composites, Kevlar/epoxy has the
highest coefficient of thermal expansion in the transverse
to the fiber direction and the lowest (negative) coefficient in the fiber direction. S-glass/epoxy has the
lowest transverse and highest longitudinal coefficients,
i.e., it is the least anisotropic thermally. In general,
especially in epoxy matrix dominated response, the coefficient of thermal expansion increases with temperature.
This trend exists also in the cryogenic range where the
coefficient approaches zero near absolute zero. In the
case of polyimide matrix composites thermal deformation
varies linearly with temperature. The moisture content
has a significant effect on thermal deformation with
some shrinkage resulting after thermal cycling at
elevated temperature. Thermal expansion is stabilized
after sufficient drying.

Residual stresses are produced during curing in angle-ply laminates as a result of anisotropic thermal deformations of the variously oriented plies. Residual strains have been measured experimentally using embedded strain gage techniques and residual stresses were computed using orthotropic stress-strain relations. Results show that, for graphite and Kevlar laminates residual stresses at room temperature are high enough to have caused damage in the plies in the transverse to the fiber direction. It has also been shown that residual stresses do not relax appreciably. The ply stacking sequence has no effect on the magnitude of average residual stresses. Residual stresses and susceptibility to cracking during curing are strongly dependent on ply layup. In laminates of $[0_2/\pm\theta_2]_s$ layup, for example, the transverse tensile residual stress at room temperature in the $\pm\theta$-deg plies increases from zero to a maximum as θ varies between 0-deg. and 90-deg. The laminate is susceptible to microcracking when this stress exceeds the transverse strength of the unidirectional lamina. Interply hybridization of laminates has only a small influence on residual stresses in the various plies.

REFERENCES

1. R.A. Schapery, "Thermal Expansion Coefficients of Composite Materials Based on Energy Principles," J. Composite Materials, Vol. 2, No. 3, July 1968, pp. 380-404.

2. B.W. Rosen and Z. Hashin, International Journal of Engineering Science, Vol. 8, 1970, pp. 157-173.

3. J.E. Ashton, J.C. Halpin and P.H. Petit, Primer on Composite Materials: Analysis, Technomic, Stamford, Conn., 1966, pp. 88-91.

4. Grumman Aerospace Corp., "Advanced Composite Wing Structures-Boron/Epoxy Design Data," Vol. II, Analytical Data, Tech. Report AL-SM-ST-8085, Nov. 1969 (AFML Contract No. F33615-68-C-1301).

5. J.C. Halpin and N.J. Pagano, "Consequences of Environmentally Induced Dilatation in Solids," in Recent Advances in Engineering Science, ed. by A.C. Eringen, Gordon and Breach, London, 1970, p. 33.

REFERENCES (Cont'd)

6.   W.T. Freeman and M.D. Campbell, "Thermal Expansion
      Characteristics of Graphite Reinforced Composite
      Materials," Composite Materials: Testing and Design
      (Second Conference), ASTM STP 497, American Society
      for Testing and Materials, 1972, pp. 121-142.

7.   N.P. Freund, "Measurement of Thermal and Mechanical
      Properties of Graphite/Epoxy Composites for
      Precision Applications," Composite Reliability,
      ASTM STP 580, American Society for Testing and
      Materials, 1975, pp. 133-145.

8.   A.S.D. Wang, R.B. Pipes and A. Ahmadi, "Thermoelastic
      Expansion of Graphite-Epoxy Unidirectional and Angle-
      Ply Composites," Composite Reliability, ASTM STP 580,
      American Society for Testing and Materials, 1975,
      pp. 574-585.

9.   I.M. Daniel, T. Liber and C.C. Chamis, "Measurement
      of Residual Strains in Boron/Epoxy and Glass/Epoxy
      Laminates," Composite Reliability, ASTM STP 580,
      American Society for Testing and Materials, 1975,
      pp. 340-351.

10.   I.M. Daniel and T. Liber, "Lamination Residual
      Stresses in Fiber Composites," IITRI Report D6073-I,
      NASA CR-134826, March 1975.

11.   I.M. Daniel and T. Liber, "Lamination Residual
      Stresses in Hybrid Composites," IITRI Report D6073-
      II, NASA CR-135085, June 1976.

12.   M.B. Kasen, "Properties of Filamentary-Reinforced
      Composites at Cryogenic Temperatures," Composite
      Reliability, ASTM STP 580, American Society for
      Testing and Materials, 1975, pp. 586-611.

13.   L.W. Toth, B.R. Lloyd and R.L. Tennant, "Determina-
      tion of the Performance of Plastic Laminates at
      Cryogenic Temperatures," ASD-TDR-62-794, Part II,
      (N64-24212), Wright-Patterson Air Force Base, Ohio,
      March 1964.

14.   C.C. Chamis, "Lamination Residual Stresses in Cross-
      Plied Fiber Composites," Proc. of 26th Annual
      Conference of SPI, Reinforced Plastics/Composites
      Division, Paper No. 17-D, Feb. 1971.

REFERENCES (Cont'd)

15. H.T. Hahn and N.J. Pagano, "Curing Stresses in
     Composite Laminates," J. Composite Materials,
     Vol. 9, 1975, pp. 91-106.

16. I.M. Daniel and T. Liber, "Measurement of Lamination
     Residual Strains in Graphite Fiber Laminates,"
     Proc. of Second International Conference on
     Mechanical Behavior of Materials, ICM-II, Boston,
     MA, 16-20 August, 1976.

17. H.T. Hahn, "Residual Stresses in Polymer Matrix
     Composite Laminates," J. Composite Materials,
     Vol. 10, Oct. 1976, pp. 266-278.

18. I.M. Daniel and T. Liber, "Relaxation of Residual
     Stresses in Angle-Ply Composite Laminates,"
     Composite Materials: The Influence of Mechanics
     of Failure on Design, Army Symposium on Solid
     Mechanics, 1976, Sept. 1976, South Yarmouth,
     Massachusetts.

19. I.M. Daniel and T. Liber, "Effect of Laminate
     Construction on Residual Stresses in Graphite/
     Polyimide Composites," Exper. Mechanics, Vol. 17,
     Jan. 1977, pp. 21-25.

20. I.M. Daniel and T. Liber, "Lamination Residual
     Strains and Stresses in Hybrid Laminates,"
     Composite Materials: Testing and Design (Fourth
     Conference), ASTM STP 617, American Society for
     Testing and Materials, 1977, pp. 331-343.

APPLICATIONS OF GRAPHITE COMPOSITES

TO DIMENSIONALLY STABLE SATELLITE STRUCTURES

JACK PRUNTY & DENNIS DUNBAR

GENERAL DYNAMICS CONVAIR DIVISION

SAN DIEGO, CALIFORNIA

ABSTRACT

This paper presents an appreciation of the success attained in the application of graphite composites to dimensionally stable satellite structures. To exemplify this, the discussion describes the performance attained in an early development article, examples of the subsequent proliferation of applications, and the accepted potential of graphite composites for the ultra-large satellite structures of the future.

INTRODUCTION

The graphite composite materials considered in this paper consist of graphite fibers embedded in thermosetting or thermoplastic matrices. Epoxy is currently the most commonly used thermosetting matrix material and polysulfone the most usual thermoplastic. For the applications treated here, the composite is made in laminated form by the successive layup of pre-impregnated unidirectional tape or pre-impregnated woven cloth. In either case, the directionality of the fibers is usually varied in steps through the thickness of the laminate to obtain the required properties.

These composites have gained wide acceptance in the aerospace industry as outstandingly efficient structural materials for a large variety of applications. Some noteworthy examples of such applications are discussed later.

In the realm of structural materials, graphite composites offer the unique capability of providing nominally zero thermal expansion or, alternatively, thermal expansion characteristics tailored to specific values. It is, of course, this attribute which has led to the adoption of these materials for a variety of dimensionally stable satellite structures. Before describing a selection of these applications, a brief discussion of some of the practical design considerations is pertinent.

In the case of theoretically zero-expansion laminates, and also in the case of those designed to exhibit a specified coefficient of thermal expansion (CTE), some deviations from the design value occur in the actual hardware. For potential applications, the attainable values are of significant interest. Tables 1 and 2 present data obtained in the Convair precision dilatometer for test coupons extracted from large structural panels. These values were attained in low-cost structure programs using routine techniques for manufacturing and material control, without attempting to control parameters specifically for the attainment of greater precision in the CTE.

These data, typical for "as manufactured" structural components, demonstrate that a CTE of 0 ± 0.10 microinch/inch °F is easily attained. This value is adequate for most applications, but could be improved to meet more stringent criteria by control of the material and processing parameters which affect the CTE.

Another significant advantage of graphite composites is given by the superior stiffness/density ratio obtainable in laminates using fibers of the high modulus or ultra-high modulus type. With a typical density of 0.060 lb/in³, laminates with properties essentially isotropic in the plane of the laminate can exhibit a value of modulus/density of 250×10^6 inches compared to a range of 100 to 112×10^6 for the structural metals: aluminum, steel, and titanium. Obviously, for stiffness-critical structures, a weight reduction approaching 60% can be obtained as compared with an equivalent metallic structure. Such significant savings have been achieved in some of the applications discussed later. In many

Table 1. HEAO-B CTE test results.

Material: GY70/X30	
Layup: (0/45/90/135) $_{2S}$	
Specimen	"Isotropic" CTE (μ in./in.°F)
	T = 68F
1	-0.050
2	-0.034
3	-0.019
4	+0.096
5	+0.064
6	-0.076
7	+0.028
8	+0.082
Avg	+0.011

Table 2. GEMS CTE test results.

Material: Modmor I/X30	
Layup: (0_3/±45/90) $_S$	
Specimen	Longitudinal CTE (μ in./in.°F)
	T = 35F
1	-0.057
2	-0.049
3	-0.003
4	-0.004
5	-0.022
6	-0.017
7	+0.003
8	-0.056
9	-0.031
Avg	-0.026

strength-critical applications, dramatic weight reductions can also be obtained with graphite composites using suitable high-strength fibers.

A characteristic of the graphite composites which must be considered for dimensionally stable applications is hygroscopy: the material readily absorbs moisture from the atmosphere. This absorption induces a dimensional change which can be as high as 20 microinches/inch in the plane of the laminate for typical manufacturing conditions and time spans. Desorption can occur in the operational vacuum environment, with an attendant reversal to the original dimensions. Some dimensionally stable instruments, such as the Space Telescope (ST) have adjustment capabilities which can accommodate this change. For other applications where adjustment is not feasible, proven techniques are available for preventing moisture absorption by sealing of the surfaces.

Finally, in concluding this brief review of the dominant design considerations, it may be noted that the nonductile nature of the material is not reflected in unduly brittle characteristics, resistance to damage is adequate for most applications, and fatigue resistance is generally excellent.

The following sections will describe some development work in support of current applications, some actual current applications, and finally some dramatic applciations planned for the near future.

APPLICATIONS

The Space Telescope

The Space Telescope (Figure 1) is a 2.4 meter (94.5 inch) diameter reflecting telescope scheduled for launch into a 500 km (311 mile) earth orbit in 1983. The project is sponsored by the National Aeronautics and Space Agency (NASA). Recently, the contract for the design of the telescope was awarded to the Perkin-Elmer Corporation, Danbury, Connecticut. The discussion here concerns development work previously accomplished by General Dynamics Convair Division to demonstrate that a graphite-epoxy structure for the telescope could satisfy the stringent dimensional stability criteria and provide structural adequacy in all respects. This technology development program was performed under contract NAS8-28201 to the George C. Marshall Space Flight Center (MSFC).

The configuration of the telescope at the time of the technology contract award is shown in Figure 1. This configuration differs from the current version, primarily in diameter. The subsequent changes, however, do not invalidate the results of previous structural development. In the MSFC contract, General Dynamics Convair designed, fabricated, and tested a half-scale model of the metering structure using Modmor I/X30 graphite-epoxy.

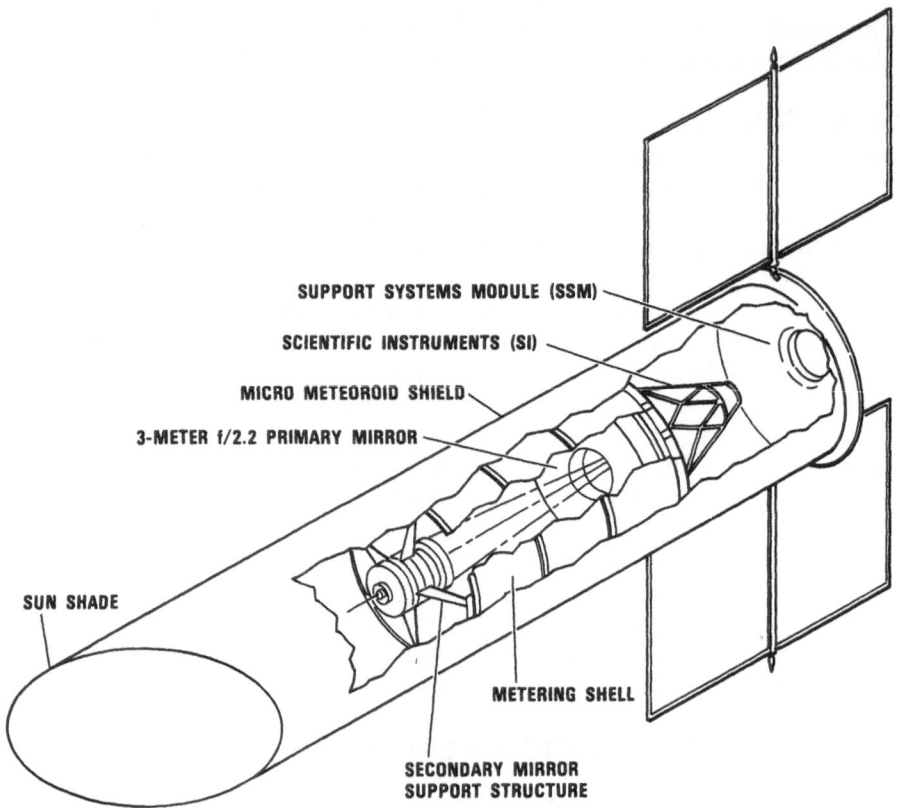

Figure 1. The Space Telescope (Phase A version).

The metering structure is visible in Figure 1 as the cylindrical "body" of the telescope between the primary and secondary mirrors. As it is seen in this view, the metering structure is enclosed in a cylindrical aluminum micrometeoroid shield. The primary function of the metering structure is to precisely maintain the spatial relationship between the primary and secondary mirrors. The secondary mirror is provided with focus, tilt, and centering adjustment but, once this adjustment is performed, it must be maintained within specified close tolerances throughout any period of observation. At the start of the technology program, the specified tolerance budget for the full-scale metering structure was:

Despace = ± 2.0 micrometers (±78.8 microinches)

Decenter = ±10.0 micrometers (±393.7 microinches)

Tilt = 4.9 microradians (1.0 arc second)

The 5.9m (232 inch) long metering structure was required to maintain the secondary mirror relative to the primary mirror within these permissible deviations during 5.55°K (10°F) structural temperature change. A little arithmetic shows that a material having a CTE not exceeding 0.034 microinch/inch°F is required to attain this dimensional stability. Structural metallics with values for the CTE ranging from 5 to 12 microinches/inch°F are completely ruled out. In contrast, as a structural material, graphite-epoxy has a unique capability of meeting these requirements.

Having established the suitability of graphite composite for this application, a study was made of the alternative shell and truss concepts for the metering structure. The conclusion of this was that both could meet the stability criteria; the truss would be lighter, but the cost of the shell structure would be lower. Since the latter criterion was the driver, with the concurrence of MSFC a decision was made to develop a shell-type structure. Figure 2 is a photograph of the half-scale metering structure which was subsequently designed and built. The GEMS (Graphite-Epoxy Metering Shell) is 2.95m (116.1 inches) long and 1.65m (64.68 inches) in diameter. As seen in the photograph, it is a simple semi-monocoque structure with a central secondary mirror support hub attached to the shell via four legs. The composite laminates were designed to exhibit theoretically zero CTE in the critical longitudinal direction. During fabrication, specimens were extracted from the shell skin panels to verify by test the attainment of an acceptable value for the CTE. As indicated in Table 2, the results were well within the allowable value of 0.034 microinch/inch°F.

Figure 2. The GEMS Metering Structure.

The completed GEMS structure was subjected to thermal-vacuum chamber tests to demonstrate the overall thermal expansion characteristics. An important aspect of this program was the demonstration of a technique for measuring microinch strains in a large structure in a practical test environment. An adaptation of the basic Hewlett-Packard laser dilatometer measurement system was developed for use under these test conditions. Figure 3 shows a schematic of this system, as used for the simultaneous measurement of the defocus and tilt modes. The laser unit was located outside the vacuum chamber to shoot through an optical quality glass window. Inside the chamber two mirrors registered on the GEMS structure, one at each of the telescope mirror reference planes. Essentially, the laser apparatus used the interferometric fringe-counting technique to measure changes in the distance between the laser optical assembly and each of the reference plane mirrors. The unit then extracted the difference between these two measurements to determine the change in the spacing between the two reference mirrors. The changes in length of the GEMS structure were then displayed in digital form and recorded on tape to a resolution of 0.5 microinch.

During the test, the GEMS was cycled between -20°F and -100°F. Since the defocus requirement was calculated to be the most stringent, the following discussion will consider this mode. Fourteen thermal cycles were made between the structural temperature extremes, and defocus measurements were obtained for each cycle. The expansion characteristics were somewhat erratic over the first few cycles, a phenomenon which also occurs with graphite-epoxy in coupon-type testing. Dilatometer coupons are routinely subjected to ten cycles to stabilize the

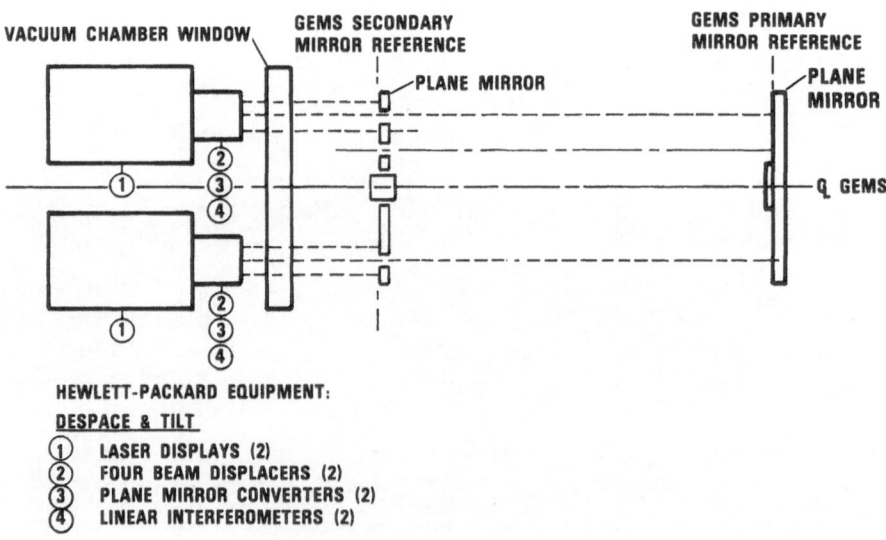

Figure 3. The Laser Dilatometer.

expansion characteristics prior to taking measurements, and complete structures are now similarly cycled prior to delivery. In the case of GEMS, a maximum value of thermal expansion of 1.88 micrometers/10°F occurred on the first cycle. A value of 1.52 micrometers/10°F was obtained on the fourteenth cycle when the characteristics had essentially stabilized.

The measured overall expansion of 1.52 micrometers/10°F exceeded the 1.00 micrometers/10°F half-scale value of the original defocus criterion. Diagnostic test runs and analyses traced the discrepancy to the canted legs of the secondary mirror support. This support geometry introduces a link-mechanism-type magnifying effect if zero CTE is not attained for the legs or for the associated support ring. An outcome of these findings was the adoption for the Space Telescope of a symmetrical support structure to correct this problem.

Considerable structural testing was also performed on the GEMS. During the test history, it was subjected to limit load some 50 times without encountering problems. Dynamic tests demonstrated a first mode fixed-base resonant frequency equivalent to 19 Hz for the full-scale metering structure. This value, which could be raised to 25 Hz by the stiffening effect of additional bolts in the attachment flange, exceeded the 15 Hz specified requirement.

Additional tests were performed with the cooperation of MSFC, in parallel with the GEMS program, to demonstrate compatibility of the structure to the orbital environment. First, the basic X30 resin system was qualified by MSFC tests to the outgassing requirements of NASA Document 40M51264. In addition, specimens were prepared which simulated the graphite-epoxy metering shell, the surrounding multi-layer insulation, and the outer aluminum micrometeoroid shell. Micrometeoroid impact tests by the Engineering Physics Branch at MSFC resulted in only roughening of the surface of the graphite-epoxy by the critical particle mass for the ST, and a two-inch diameter hole with only peripheral delamination by three times the critical particle mass.

The rather comprehensive testing in support of the ST did much to promote the use of graphite-epoxy for satellite structures, and lead directly to other applications described in the following sections.

HEAO-B Structures

The HEAO-B is the High Energy Astronomy Observatory, B Mission, an x-ray experiment scheduled for earth-orbital operation in 1978. Figure 4 depicts the HEAO-B satellite, which carries a powerful x-ray telescope coupled to a comprehensive array of instruments. The experiment is designed for detailed analysis of a great number and variety of galactic and extra-galactic x-ray objects. The forward end of the telescope can be seen in the figure as the large ellipse, surrounded by the three aspect sensors and the monitor proportional counter.

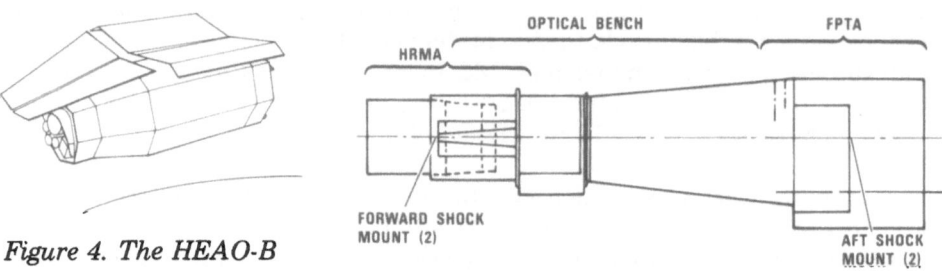

Figure 4. The HEAO-B satellite.

Figure 5. The HEAO-B telescope.

Figure 5 shows the general arrangement of the telescope. Attached to the forward end of the optical bench is the High Resolution Mirror Assembly (HRMA), a grazing incidence collector composed of concentric fuzed quartz cylinders. At the aft end, the optical bench is attached to the Focal Plane Transporter Assembly (FPTA) which houses a rotating array of instruments.

Graphite-epoxy was selected as the most suitable material for the optical bench and for the backbone structure of the HRMA. These components, which are described in the following sections, were designed and fabricated by General Dynamics Convair Division under subcontract to American Science and Engineering of Cambridge, Massachusetts.

HEAO-B OPTICAL BENCH — The completed optical bench is shown in Figure 6. It is a ring-supported, semi-monocoque structure, 119.12 inches long and 41.2 inches maximum diameter. The shell of the bench is constructed from four laminated panels joined by splice straps bonded with EA934 room temperature curing adhesive.

Figure 6. The HEAO-B optical bench.

The dominant design criteria, the first of which dictated the use of graphite-epoxy were:

- CTE longitudinally and circumferentially = 0±0.10 microinch/inch°F
- Shell thickness — modulus product in both directions = 1.5 x 10⁶ lb/in

These criteria were satisfied by a 0.104-inch thick, 16 ply, GY70/X30 laminate of $(0/45/90/135)_{2S}$ configuration. Ultra-high modulus GY70 fiber was used to meet the stiffness criterion for minimum weight and to attain a CTE isotropic in the plane of the laminate, within the specified range. The X30 resin was selected on the basis of freedom from microcracking, a phenomenon which can drastically affect thermal expansion characteristics.

It is noteworthy that, even with a laminate designed primarily for nominally zero thermal expansion, an elastic modulus 50% higher than the value for aluminum was attained with a material 36% less dense. This indicates that the weight of the basic shell structure is 43% of the weight of an aluminum structure designed to the same stiffness criterion. Compliance with the stiffness criterion was demonstrated by tests of three coupons from each of the four skin panels. The test values for the tensile modulus ranged from 15 to 17 x 10⁶ psi. As shown in Table 1, compliance with the specified CTE was also demonstrated by dilatometer tests of a longitudinal and transverse coupon extracted from each of the four skin panels.

A feature of interest in this application was the adherence to stringent dimensional tolerances on the optical bench assembly. Exemplifying this are the three aspect sensor mounts, conspicuous in Figure 6, which were maintained within ±0.005 inch relative to the forward face of the bench and the optical axis.

An additional point of interest is that the NASA permitted a protoflight approach for the optical bench. In this approach to the design of flight hardware, a factor of safety = 3 is used, structural integrity is demonstrated by analysis only, and all requirements for structural testing are eliminated. This curcumstance demonstrates the high degree of confidence which has been established for this type of graphite-composite structure.

HRMA CYLINDERS — These structural components, which support the concentric fuzed quartz cylinders in the mirror assembly, are shown in Figure 7. The cylinders vary in size from 10.740 inches diameter by 20.955 inches long to 25.410 inches diameter by 20.615 inches long.

Four prime design requirements for the cylinders were decreed to minimize distortion in the mirror assembly:

1. Attain a CTE of 0.3 microinch/inch°F longitudinally and circumferentially for compatibility with the thermal expansion of the fuzed quartz mirrors.

Figure 7. The HRMA cylinders.

2.   Comply with specified shear and bending stiffness requirements.

3.   Eliminate dimensional changes due to moisture.

4.   Maintain geometrical tolerances on the outer surface of ±0.005 inch on diameter and within 0.003 inch concentricity.

The CTE requirement could have been met with LR-35, a version of INVAR low-expansion nickel alloy. A comparison of the modulus/density ratios, however, indicates that INVAR cylinders would weigh 2.4 times as much as graphite-epoxy versions for equal stiffness. As in the case of the optical bench, ultra-high modulus GY-70/X30 graphite epoxy was selected to enable compliance with the stiffness requirements for minimum weight. The laminates evolved by the analysis were 0.572 inch thick, 88 ply, $(0_8/90_{12}/\pm50_{12})_S$ for the outer cylinders, and 0.286 inch thick, 44 ply, $(0_4/90_6/\pm50_6)_S$ for the inner cylinders. (This simplified notation does not indicate the actual stacking sequence of the plies.)

As shown in Figure 7, the cylinders were encapsulated in aluminum foil to inhibit moisture effects. Actually, a simple deposition coating is available for this purpose, but at the time of the cylinder fabrication the facility was not of adequate size. The technique adopted had been previously demonstrated as a means of eliminating moisture by long term dilatometer testing for moisture-induced dimensional changes. This sealing technique consists of applying, by adhesive bonding, a double layer of 0.0007-inch-thick aluminum foil with a schedule of initial, intermediate, and final dryout cycles.

Close control of the outer surfaces was attained by curing the cylinders in female molds with a tolerance of ±0.003 inch on the mold diameters. The precision molding technique developed for this purpose is now finding a variety of applications. Also, the general development of this type of graphite/composite cylinder creates the possibility of providing lightweight substrates for the mirror cylinders of future very large x-ray telescopes.

The preceding examples give an appreciation of the successful exploitation of graphite-epoxy to meet demanding requirements in current applications. The following sections describe some of the more exciting and challenging applications planned for the future.

Large Space Structures

With the advent of Space Shuttle, space operations are entering a period of transition from single missions and limited-duration activities to continuous operations embracing a wide spectrum of applications. A new dimension in space structures is being introduced with extremely large, ultra-low-density arrangements required for such applications as satellite power systems, phased-array antennas, microwave radiometry arrays, and large platforms providing common structural and operational support for a variety of experiments and operations. A topically pertinent example is the solar power satellite (SPS) concept shown in Figure 8, an immense satellite structure which can significantly augment electrical energy supplies at a time when many terrestrial energy sources are approaching depletion. Such enormous structures require assembly in space because of payload volume and weight limitations of the Shuttle and even of future heavy lift launch vehicles. Graphite composites are expected to play a dominant role in this challenging field. General Dynamics Convair is engaged in development in this area under contract to the NASA Johnson Space Center and the Marshall Space Flight Center.

Critical considerations for such large orbital arrays are thermal and dynamic response. Cumulative distortions due to thermal gradients in large arrays could cause major transient geometric changes that would degrade the performance of space systems, and initially create acute problems in the in-space assembly of ultra-large structures. The magnitude of the potential thermal deformation problem is demonstrated in Figure 9. This figure shows, for various materials and surface characteristics, the thermal deformation of a 200m long space structure beam with two beam caps exposed to solar heating and one cap shadowed. The advantage of graphite composite materials for thermal distortion control is evident.

NASA and industry are studying the problems involved in transferring such large satellite structures from the ground to orbit. Two concepts showing promise are described here. Each has a role to play, depending on the ultimate size of the structure and the time frame when the structure is required.

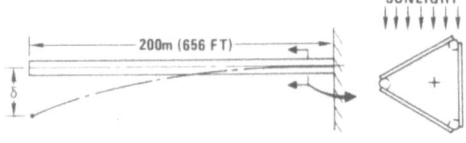

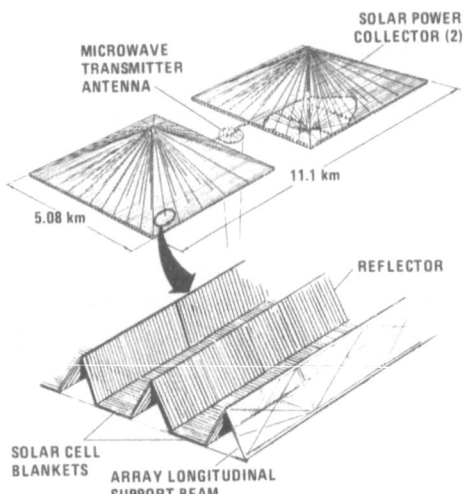

MATERIAL	SURFACE	CTE (PPM/°F)	ΔT MAX (°F)	TIP DEFLECTION δ (m)	(IN.)
ALUMINUM	WHITE	12.00	103	16.45	647.6
GRAPHITE (ISOTROPIC)					
AS/P-1700	WHITE	1.42	103	1.95	76.8
AS/P-1700	BARE	1.42	252	4.78	188.0
GY70/P-1700	WHITE	0.10	103	0.14	5.0
GY70/P-1700	BARE	0.10	252	0.32	11.8

Figure 8. The Solar Power Satellite.   *Figure 9. Graphite composites eliminate critical distortion problems.*

1. Deployable Structures — Folded and packaged modular elements can be carried in the shuttle, then deployed into approximately 100 meter building blocks. Such building blocks can be assembled in orbit into larger structures.

2. In-space Fabrication of Structures — Automated fabrication of very large structures from raw tape material can be accomplished in orbit using a "beam builder" machine. This concept is geared for the longer term, kilometer size, structures such as Solar Power Satellites.

DEPLOYABLE STRUCTURES — The deployable structure concept is based on a fundamental geometric shape, the tetrahedral truss, which has been so designed as to be foldable into a closely packaged element. The deployable tetrahedral truss was the outcome of studies performed at General Dynamics Convair during the late 1960s. Since that time, concept development has continued through several major NASA study contracts. In 1970, General Dynamics was awarded the patent on the concept with a provision for full usage rights by NASA.

When designed for large space strucures to be carried up in the Shuttle, the deployable tetrahedral truss is packaged as a 4.4 meter diameter module 6 meters long. Three such packages can be carried in the Shuttle as shown in Figure 10. Each package deploys into a hexagonal structure 335 feet by 290 feet. These structures can be joined in orbit into a single large satellite. Even larger structures can be assembled, using this modular approach with multiple shuttle launches. Launch costs, however, suggest that transporting material in a denser form — allowing more structure per launch, is desirable for extremely large structures such as Solar Power Satellites. This consideration resulted in the concept of in-space fabrication.

IN-SPACE FABRICATION — This concept of manufacturing structural beams in orbit will be exploited for many and various applications. These will range from moderately large sized structures to the tens of square kilometers size category required for the Solar Power Satellite.

At the heart of this concept is the beam building machine, which will automatically fabricate a basic structural beam from reels of preconsolidated graphite-thermoplastic strips. This basic beam, shown in Figure 11, is of triangular cross-section and open-truss configuration. A semi-schematic of the beam builder is presented in Figure 12. As shown in this figure, the graphite-thermoplastic strip material is supplied to the machine on reels. The feasibility of using the beam builder to fabricate very large structures is illuminated by the fact that one 14-foot diameter reel, as sized to the constraints of the Space Shuttle payload bay, will hold sufficient strip material for 17 miles of a structural element! A total of six reels is carried by the beam builder — one for each of the three cap elements and one for each of the three side-bracing arrangements. In the fabrication process, the strip material for each structural element is first heated to the forming temperature, then rolled to the required cross-section and subsequently cooled to regain rigidity prior to the assembly procedure. In the assembly stage, the side bracing members are cut off, automatically manipulated into position relative to the beam caps, and then attached to the caps by ultrasonic welding of the thermoplastic material. The basic forming process, a patent of General Dynamics Convair Division, is known as Rolltrusion.

Future ultra-large, very low density structures can utilize the basic truss beam for the caps and bracing members of beams an order of magnitude greater in size. Development of in-space fabrication techniques is being vigorously pur-

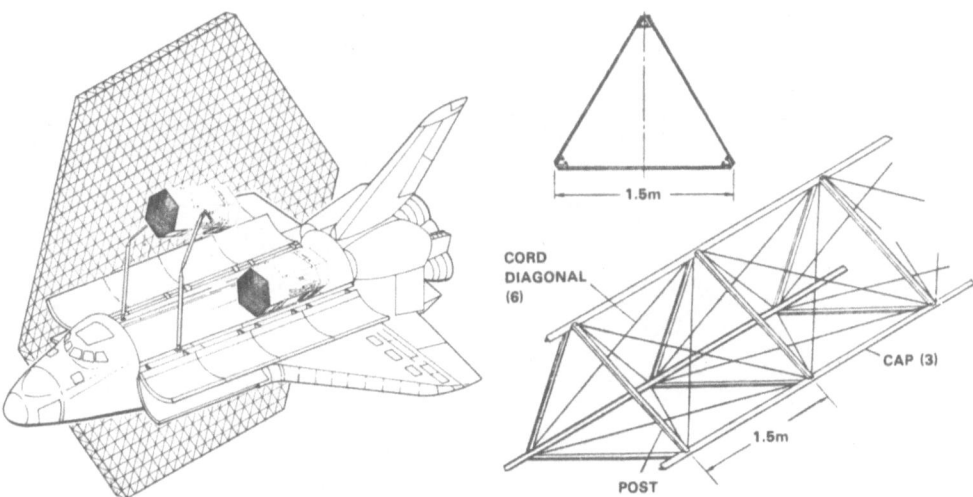

Figure 10. Deployable structure modules in the Shuttle.

Figure 11. Beams employ lightweight, open-truss construction.

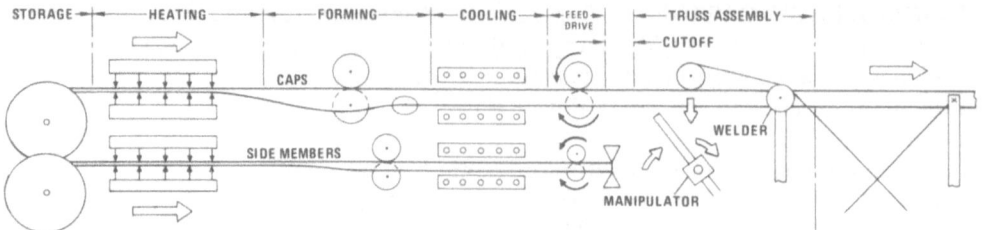

Figure 12. Semi-schematic of the beam builder (typical for three places).

sued by NASA and industry, and satellites of impressive dimensions should be feasible in the near future. To illustrate this, Figure 13 depicts a pioneering application of the beam builder. In this case, a single Space Shuttle mission, planned by the NASA JSC for 1982, will fabricate and assemble a 200-meter long satellite platform. This accomplishment will introduce an era of satellites of very large size, and will pave the way to the ultra-large orbital structures of the not too distant future.

Figure 13. A single Space Shuttle mission fabricates and assembles a 200-meter long platform.

CONCLUSIONS

In conclusion, the application of graphite composite to large satellite structures is essentially current state-of-the-art. The adoption of composites for planned spacecraft structures can realize significant benefits in stiffness-critical and strength-critical applications and can be a virtual necessity where extreme dimensional stability is required. For the future, the unique characteristics of these high performance materials can offer great benefits to large space structure programs by contributing substantially to the minimization of total program costs and to the maximization of operational efficiency.

REFERENCES

1. CASD-NAS-75-015, *Design, Fabrication, and Test of a Graphite/Epoxy Metering Shell (GEMS) Final Report,* J. Prunty et al, April 1975.

2. *The HEAO-B Optical Bench,* W.D. Antrim, R.L. Hall, R. Larson, and A. DeFuria, SAMPE, Volume 20, 1975.

3. *The Application of Thermally Stable Graphite Composites to a Space Telescope Structure,* J. Prunty and R.L. Hall, SAMPE, Volume 22, April 1977.

4. CASD-HEA-76-002, *The HEAO-B Optical Bench and HRMA Support Cylinders, Final Report,* J. Prunty, et al, June 1976.

ACKNOWLEDGEMENTS

The Graphite-Epoxy Metering Shell (GEMS) program was performed under NASA MSFC contract NAS8-28201. The Contracting Officer's Representative was Mr. Carl Loy. The HEAO-B Optical Bench and HRMA Cylinders Program was accomplished under subcontract to American Science and Engineering, Inc. Mr. R.L. Hall was project engineer for the telescope. HEAO-B is a NASA project managed by the Marshall Space Flight Center. The Principal Investigator is Dr. Riccardo Giacconi. Scientific direction is by a consortium of institutions, including:

The Smithsonian Astrophysical Observatory
The Massachusetts Institute of Technology
The Goddard Space Flight Center
The Columbia Astrophysics Laboratory

The experiment prime contractor is American Science and Engineering, Inc. The spacecraft contractor is TRW. The Space Construction Automated Fabrication Experiment definition study is under Contract NAS9-15310 to NASA JSC. Mr. Lyle Jenkins is the Contracting Officer's Representative.

EXPANSIVITY CHARACTERIZATION OF A GRAPHITE-REINFORCED ADVANCED COMPOSITE

A. T. Calimbas and T. J. DeLacy

Ford Aerospace & Communications Corporation
Western Development Laboratories Division
Palo Alto, California

ABSTRACT

Graphite-reinforced advanced composites are used extensively in communication satellites because of their excellent strength-to-weight ratios and exceptional dimensional stability. These composites pose unique challenges, however, in both preconditioning and measurement techniques which must consider the behavior of complex structures and the stabilization of strain energy in very directional materials. This paper describes the experimental studies made to correlate the expansion characteristics of a graphite-fiber/epoxy-resin matrix material with thermal cycling effects. Nondestructive testing was employed to identify the type and extent of physical change indicating probable mechanisms of stress relief in the material during cyclic thermal loading. It is shown that the expansion characteristic of the composite is influenced by sample thermal history and by relaxation of stresses induced during manufacturing. Techniques stabilizing composite expansivity are discussed.

INTRODUCTION

In recent years the application of composites - more pointedly, graphite-fiber-reinforced plastics - in space technology has drawn major attention because of two exciting properties: high strength-to-weight ratios and excellent dimensional stability. Notable examples of advanced spacecraft applications include:

- The cargo bay doors on the Space Shuttle, which are made of graphite-epoxy (G/E), because of the excellent strength-to-

weight ratio exhibited by the composite. High-modulus
properties of the fibers also provide the stiffness necessary
for a working platform.

- G/E antennas featured on both the Viking and Mariner
  Jupiter-Saturn spacecraft. The parabolic surface accuracies
  of the antennas are maintained throughout the wide range of
  thermal environments encountered by exploratory spacecraft.
  Long-term stability is achievable through the use of
  advanced composites.

- The metering truss on the large space telescope (LST).
  Lightweight properties, stability, and stiffness provide
  important advantages of G/E in this key exploration
  development.

The trend in the space industry is toward greater utilization
of G/E. For example, the next generation of international telecom-
munications satellites will use G/E not only in structure and
antennas but also in filters, feeds, and many other microwave
components.

The design and application of a dimensionally stable
advanced (composite) antenna systems require characterization of
materials whose properties are highly directional and anisotropic.
In addition, constituents of the composite typically exhibit
disparate properties (mechanical and thermal). Graphite fibers by
themselves have a small, axial negative coefficient of thermal
expansion (CTE); the epoxy matrix that binds the composite together
typically has a large, positive coefficient of thermal expansion.
The combined CTE for the composite becomes a function of not only
fiber and matrix expansivity characteristics, but also, to a first
order, fiber plies and orientation. Micromechanical analyses of the
composite yield expressions for the composite directional
coefficient of expansion with the following terms [1]:

$$\{\alpha_{ply}\} = \begin{Bmatrix} \alpha_{ply\ x-x} \\ \alpha_{ply\ y-y} \\ \alpha_{ply\ z-z} \end{Bmatrix}$$

$$= \left\{ \begin{pmatrix} semiempirical \\ constants \end{pmatrix} \begin{pmatrix} fiber\ coefficient \\ of\ expansion \end{pmatrix} \begin{pmatrix} actual\ fiber \\ volume\ ratio \end{pmatrix} \right.$$

$$\left. + \begin{pmatrix} semiempirical \\ constants \end{pmatrix} \begin{pmatrix} matrix\ coefficient \\ of\ expansion \end{pmatrix} \begin{pmatrix} actual\ matrix \\ volume\ ratio \end{pmatrix} \right\}$$

where $\begin{Bmatrix} \alpha_{\text{ply x-x}} \\ \alpha_{\text{ply y-y}} \\ \alpha_{\text{ply z-z}} \end{Bmatrix}$ is a directional matrix.

The above expression is based on assumptions which make quantitative predictions stray from experimentally measured values. This is not to denigrate well established analytical models, but rather to demonstrate that certain parameters affect the behavior of a composite and that these parameters are not likely to be embodied in a single model. Specifically, fiber orientations within a given composite ply, cross-sectional properties and distribution of individual fibers, resin voids, thermal history (strain), etc, are all parameters that contribute to variance between predicted and actual properties. While one can establish equations to fit a given set of measurements, the process must be repeated for each unique sample.

The need for increased stability and application of advanced composites for use in space requires that measurement techniques be developed for vehicle (satellite) life expectancy heretofore unprecedented. Accordingly, instrumentation development requires a thorough understanding of both static and dynamic behavior that influences the accuracy of measurements as small as 0.5×10^{-6} meter. The following sections describe studies undertaken by Ford Aerospace & Communications Corporation, Western Development Laboratories Division, to provide the understanding necessary for the development of repeatable design data and measurement hardware. Essentially these studies were an embryonic effort to understand stability mechanisms in advanced graphite composites.

EXPERIMENTAL TECHNIQUES

Expansivity work at FACC has centered on telemicroscopes as the sample length change detector, their major advantage being a contactless method of monitoring the sample expansivity behavior with temperature changes. Figure 1 shows the basic experimental configuration used for the work. Salient features of the equipment include:

- All measurements are made in partial vacuum of 2 to 10 torr to minimize absorbed water effects following drying of the sample.

- The thermal chamber surrounds the sample completely, with compensating heating and cooling at the window apertures to minimize perturbations of the sample temperature field.

- The telemicroscopes have a power of 50 with a least count of 0.5 micrometer.

- Automatic temperature controllers are used.

- Thermocouple temperature sensors are monitored on a digital readout and recorded on a multipoint, strip-chart recorder.

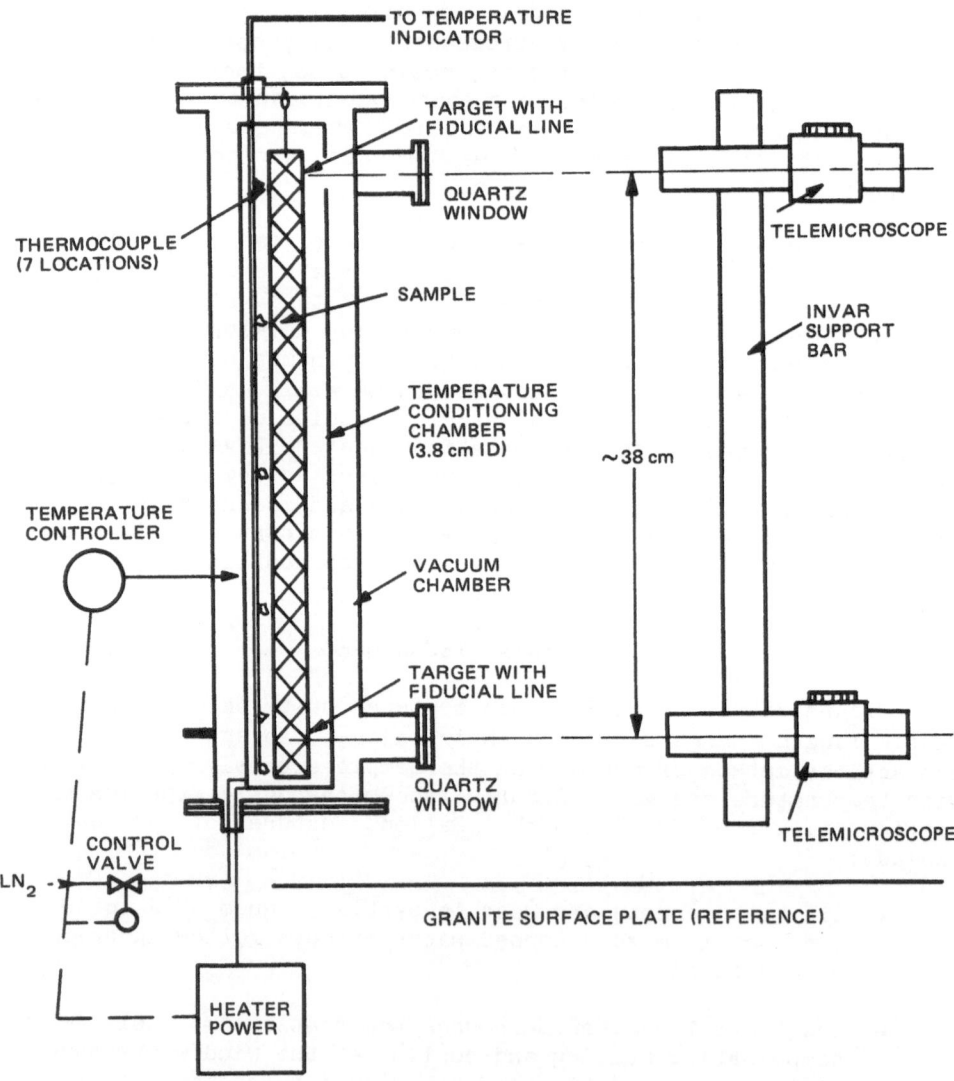

Figure 1. Schematic of Existing Thermal Expansion Measuring
Equipment

An uncertainty analysis [2] of the experimental configuration yielded the results graphically shown in Figure 2. In brief, using two telemicroscopes and a $\pm$ 5°C uncertainty in the sample temperature, the coefficient of expansion determinations were accurate to $\pm$15%.

As work progressed, it was noted that a significant improvement in the measurement repeatability and a reduction of the measurement uncertainties could be achieved by monitoring the sample length changes relative to a fused silica standard whose expansivity was calibrated by the National Bureau of Standards. As such, all measurements of sample expansivities are now made relative to the NBS standard material. The uncertainties of this technique are $\pm$10%.

The method by which the sample and NBS standard are suspended in the measurement furnace is depicted in Figure 3. Precision tools and jigs are used to assure that the hole (oversize) in the tubular sample is perpendicular to its centerline. Small weights are hung

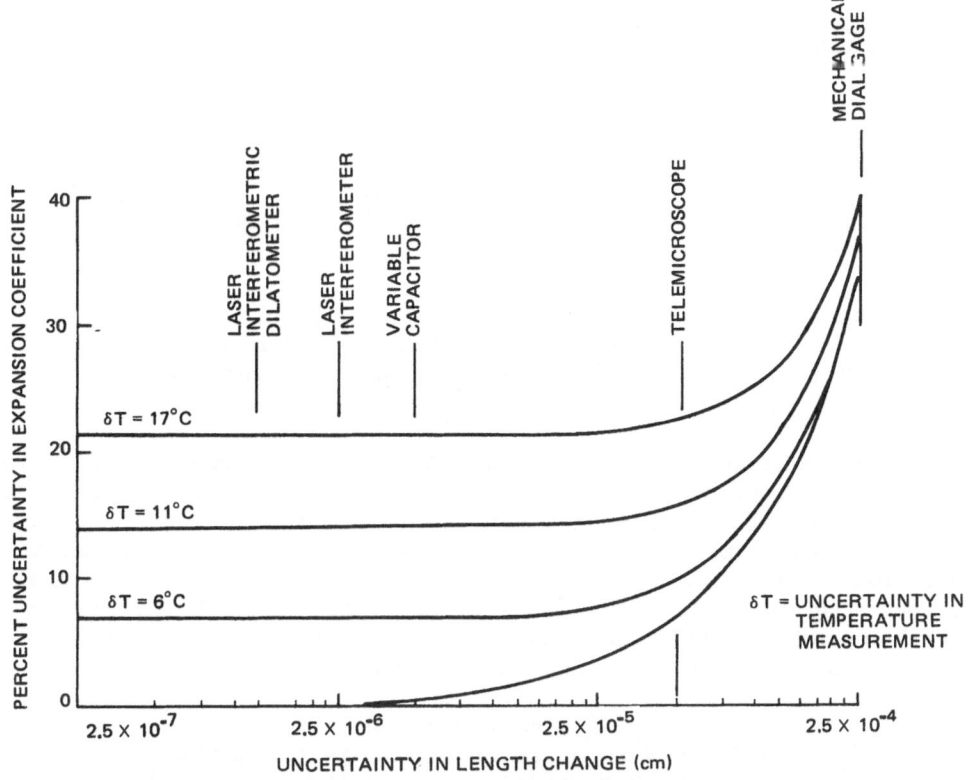

Figure 2. Experimental Measurement Uncertainties

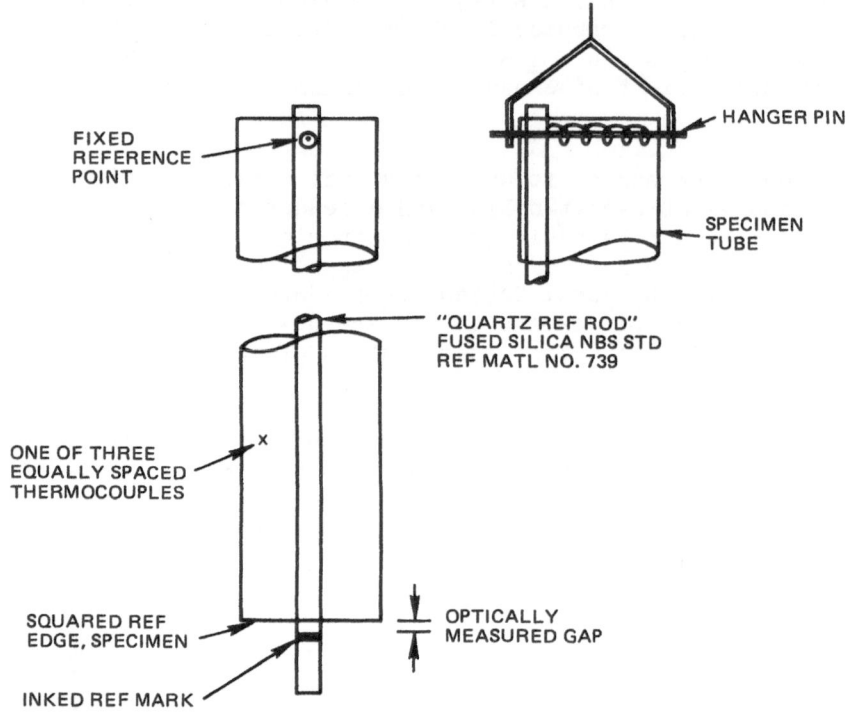

Figure 3. CTE Test Configuration (Specimen Hung in Evacuated
Chamber with Provisions for Radiant Heating and Cooling)

around the periphery of the tube so that the sample and reference
rod hang vertically in the chamber. All adjustments are made on the
surface plate ($\pm 6.25 \times 10^{-7}$ m) on which these measurements are
performed.

The technique employed for expansivity measurements at FACC is
more or less tailored for tubular samples. This sample geometry has
been found to eliminate significant error due to bending or twisting
motions in the sample. We have, on occasion, measured samples with
geometries other than tubes (ie, flat sheet, honeycomb cored, etc),
but have had to use special precautions to assure that the sample
length change was not unduly influenced by bending or twisting. The
measurement procedure included the following significant steps:

1. Prior to expansivity measurements, the sample was outgassed
    at 120°C for 4 hours under vacuum in the measurement
    chamber. By thermogravimetric analyses (TGA), it was
    determined that these outgassing conditions removed

measurable (~85%) volatiles in the samples tested. Because of the spacecraft application for which these measurements were made, it was intended to study the material (G/E) in its orbital condition (eg, moisture-free condition), thus eliminating one of the parameters thought to influence G/E expansivity.

2. The sample was thermally cycled three times over the full temperature range of the measurements. Studies by FACC and others [3] have shown that advanced composites with locked-in stresses exhibit a rapidly changing expansivity as the sample undergoes initial temperature cycling. Using non-destructive evaluation [4], evidence was found that if macrocracking or disbond was the major stress relief mechanism, its manifestation is generally exhibited within the first three to five temperature cycles. Further, the measurement equipment stabilizes and enhances the repeatability of the measurement.

3. Baseline positions of fiducial lines were taken at 20°C.

4. Using the telemicroscopes, the sample length changes were measured as the sample came to equilibrium at each measurement temperature.

Rapid temperature cycling was performed in the gas-driven system shown in Figure 4. With our interest in the expansivity of these composites after thousands of thermal cycles, a convective heat transfer system was selected to allow testing to be accomplished in weeks rather than years. It should be noted that only dry GN_2 is circulated in the chamber to reduce moisture effects.

The ultrasonic transmission-reflection technique which was used to detect and monitor macrochange in the samples is shown in Figure 5. Ultrasonic energy is transmitted through the test specimen (tubular element) to a reflector bar around which the tube and standard cylinder are mounted. If disbond is present, reflection of the sound beam at the interface results in less transmitted energy; the loss is detected as a decrease in the strength (amplitude) of the reflected signal. Figure 6 shows the signal display on the cathode ray tube of the ultrasonic instrument. The gated signal, input to a recording amplifier, provides an output voltage of the proper level to the recorder.

Figure 7 shows the arrangement of the ultrasonic instrumentation. Referring to the figure, a recorder is mechanically linked to scan and index motions of the transducer and is electrically connected to the recorder amplifier. With a signal in the gate, the recorder receives the amplified signal and "writes" during that time. A series of line scans synchronized with the movement of the

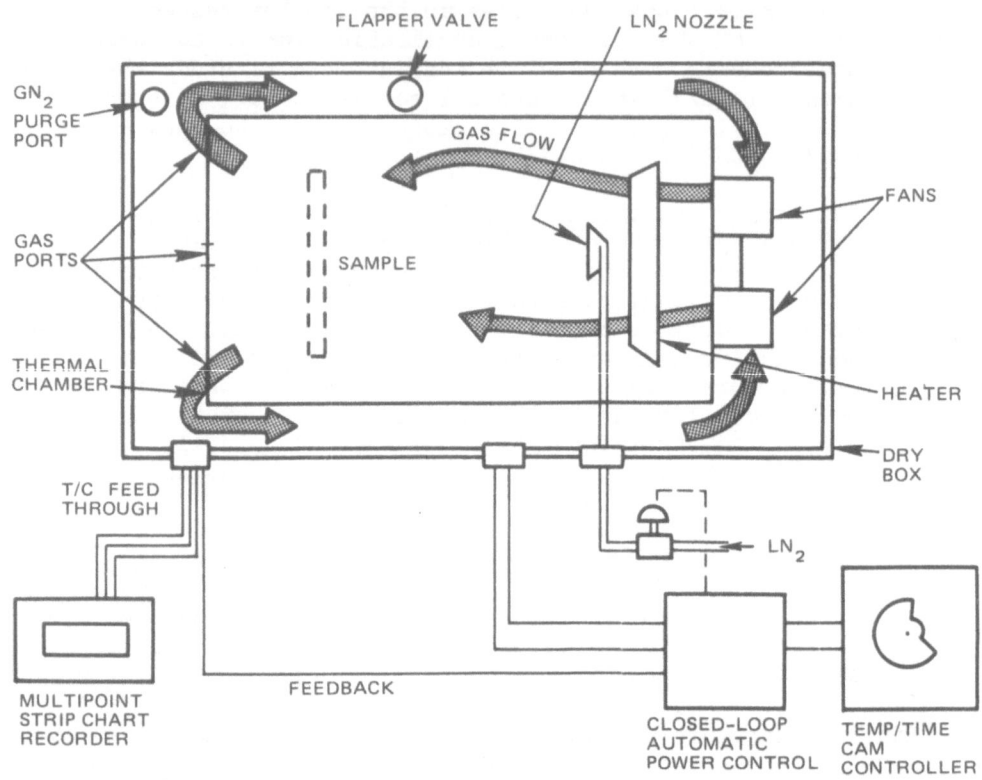

Figure 4.   Schematic of Gas Cycling System

transducer provides a full-scale plan view of the tube showing
defects in either black and white or variable shades.

     The standard used in obtaining the ultrasonic measurement is
shown in Figure 8. The test specimen is centered in an ultra-
sonically translucent (minimum attenuation) cylinder which contains
sprockets for centering the specimen. Lead tape is placed on
the outside of the cylinder to obtain a dampening pattern which
represents the threshold (information index) for detecting and
monitoring change in the specimens. Selection of the technique
was based on the resolution and repeatability for equivalent
test setups involving graphite-epoxy composites. Providing a
2:1 magnification of circumferential change, the standard facili-
tates means for identifying accountable defects and for comparing
results for flaw growth data collection aimed at measuring thermal
stability during cyclic testing. The sensitivity range used
for these measurements was ±2 dB.

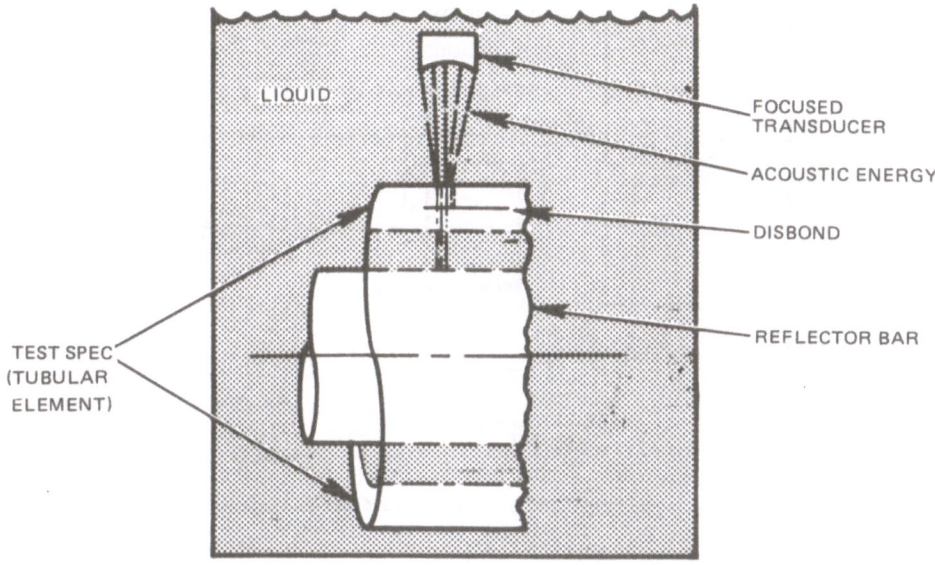

Figure 5.   Ultrasonic Transmission-Reflection Test Method

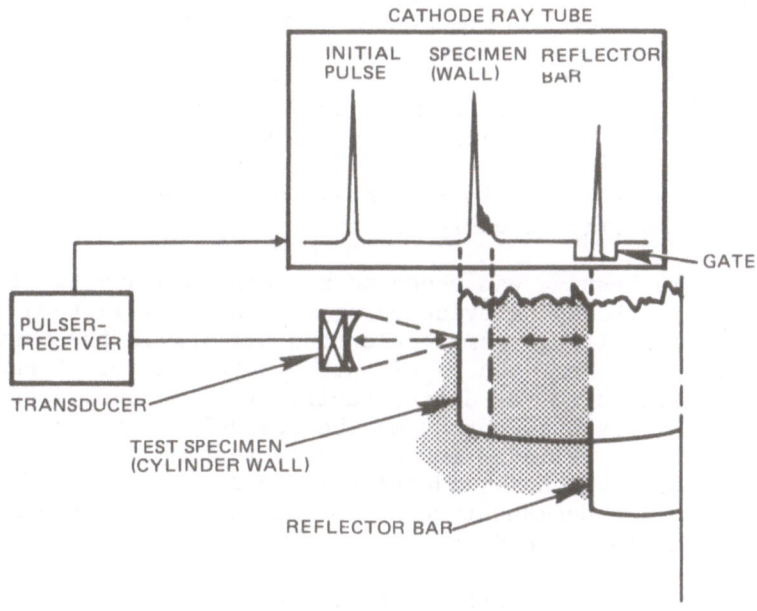

Figure 6.   Ultrasonic Transmission Signal Display Using Reflector
Technique

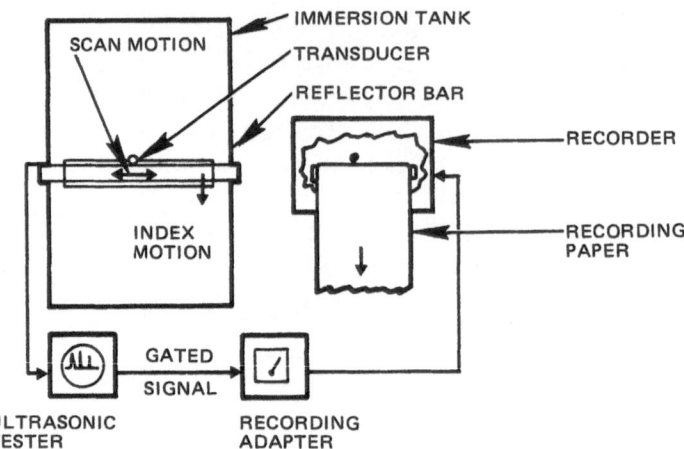

Figure 7. Ultrasonic Test System Arrangement

RESULTS AND DISCUSSION

Results of our studies show that changes in expansivity are
predisposed by thermal history (ie, cyclic exposure) as well as
stress buildup induced during manufacturing. Factors influencing
dimensional stability have been reported by independent
investigators [1,3,5,6]. These factors include viscoelastic
effects, creep, and stress relaxation mechanisms relating to material
properties and test techniques. For the subject study, cyclic
thermal conditioning was used to relieve internal stresses that were
shown to produce a change in mechanical properties and/or
dimensional stability in early testing.

Ultrasonic testing was shown to be complementary to CTE
measurements in providing evidence of internal damage (cracks,
disbond, etc) resulting from accelerated testing in the service
environment. Moreover, use of CTE without the aid of ultrasonic
data can be misleading for design verification in that mechanical
properties of a given component may be severely degraded as a
result of stresses relieved during cyclic testing. In addition
to providing data which may identify induced damage, ultrasonic
testing provides a permanent record of macrodefects (or changes)
in the material for both diagnostics and quality assurance.

Figure 9 shows the expansivity curve, $\Delta \ell / \ell$ vs temperature, for
a typical graphite-epoxy tubular element. The specimen, a 10-ply,
hybrid composite comprised of HMS and GY-70 laminates, shows near
zero expansion (α = 0.27 x 10^{-6}/°C over a reported temperature

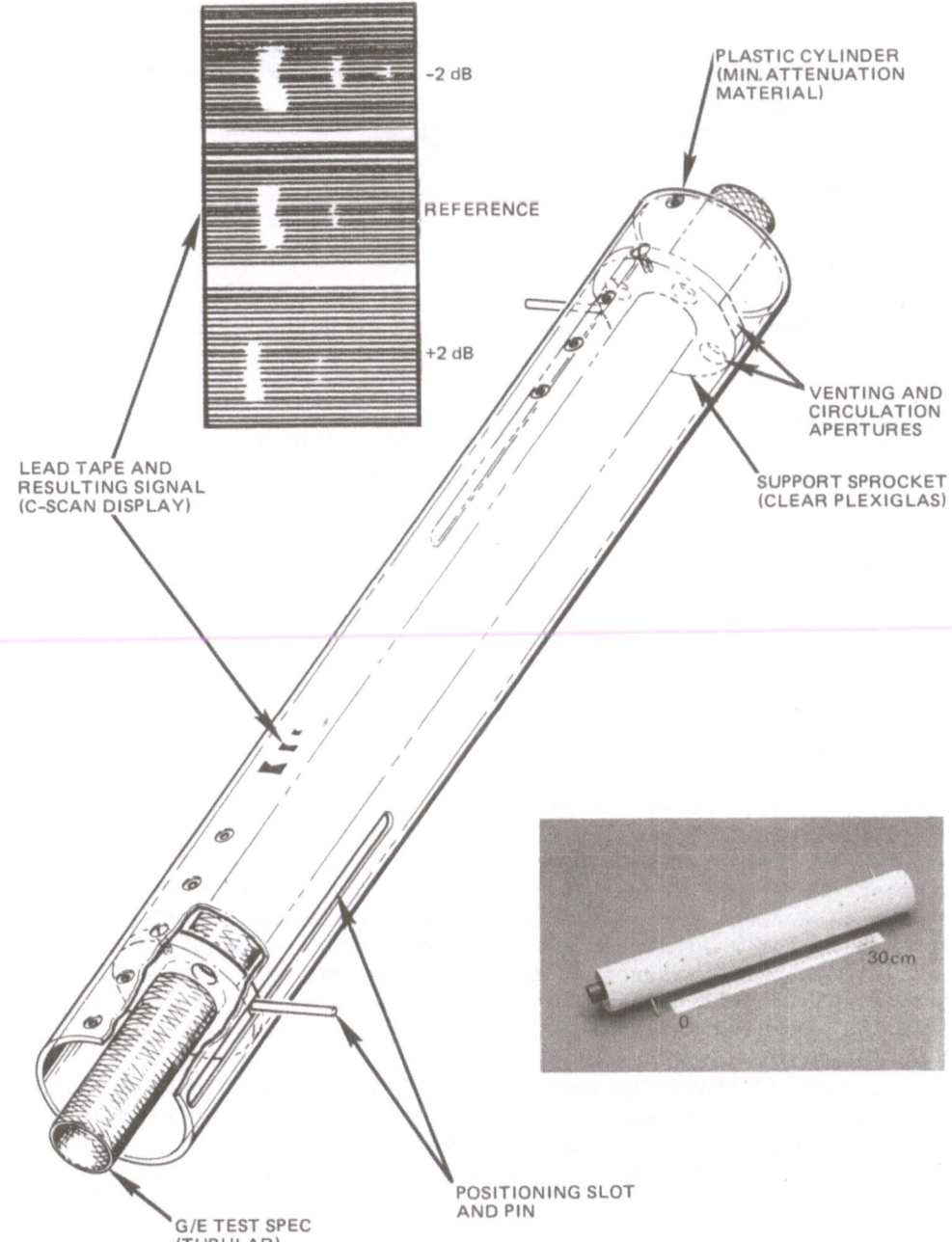

−2 dB

REFERENCE

+2 dB

PLASTIC CYLINDER
(MIN. ATTENUATION
MATERIAL)

VENTING AND
CIRCULATION
APERTURES

SUPPORT SPROCKET
(CLEAR PLEXIGLAS)

LEAD TAPE AND
RESULTING SIGNAL
(C-SCAN DISPLAY)

30 cm

0

POSITIONING SLOT
AND PIN

G/E TEST SPEC
(TUBULAR)

Figure 8.   Ultrasonic Test Standard for Measuring Macrochange

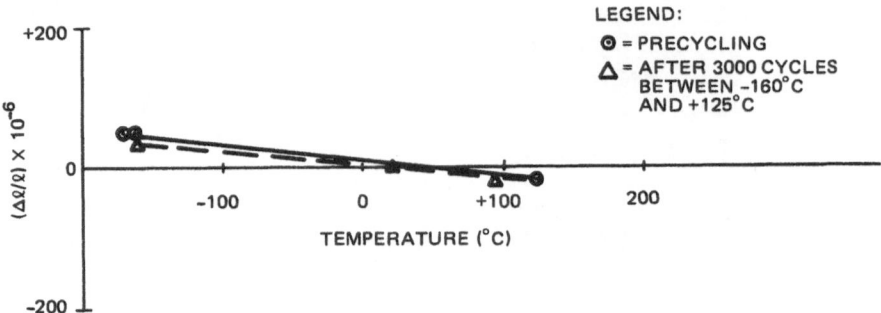

Figure 9.   Expansivity Data Before and After 3000 Temperature Cycles

range between plus and minus 120°C. The change in alpha ($\Delta\alpha$)
may be seen to be in the order of $\lesssim 0.07$ x 10^{-6}/°C during simulated
service, ie, up to 3000 life cycles. Ultrasonic C-data obtained
in conjunction with these tests verified the stability of the
component (bonded laminate) following thermal cycling.

Where tooling and/or other manufacturing constraints induce
internal stress, dissipation usually results in a change in CTE
response during early cyclic testing. Studies by FACC have shown
this response to be frequently accompanied by inelastic behavior
which may be monitored (detected) by ultrasonic transmission-
reflection measurements sensitive to bond interfaces (disbond)
and equivalent crack planes. Generally, dissipation of stresses
is ordered and results in microcracking, minor disbond, or combination
mechanisms which do not materially affect mechanical properties.
Further, laboratory studies (Figure 10) show that repeatable CTE
may be correlated with defect growth (stability) in instances
where macrodefects are detected and monitored during thermal
cycling. Accordingly, the argument of complementary test data
to measure internal stress damage which cannot be detected by
conventional CTE measurement must be advanced. For example,
stress buildup (or dissipation) may be studied with the aid of
acoustic emission analysis for further understanding.

Figure 11 shows the results of CTE measurement in which slope
reversal may be seen during conditioning cycling of an experimental
(hybrid) composite. Ultrasonic C-scan data (Figure 12) show the
condition of the specimen before and after thermal cycling.
Extensive cracks and delamination were nucleated during the first
few cycles. While erratic CTE measurement is likely to be
accompanied by severe macroevents in the material, the degree and
mechanisms of stress relief are not currently predictable during
early testing. Studies by FACC have shown abrupt changes in $\Delta\ell/\ell$
following as many as 20 temperature excursion cycles (+120°,

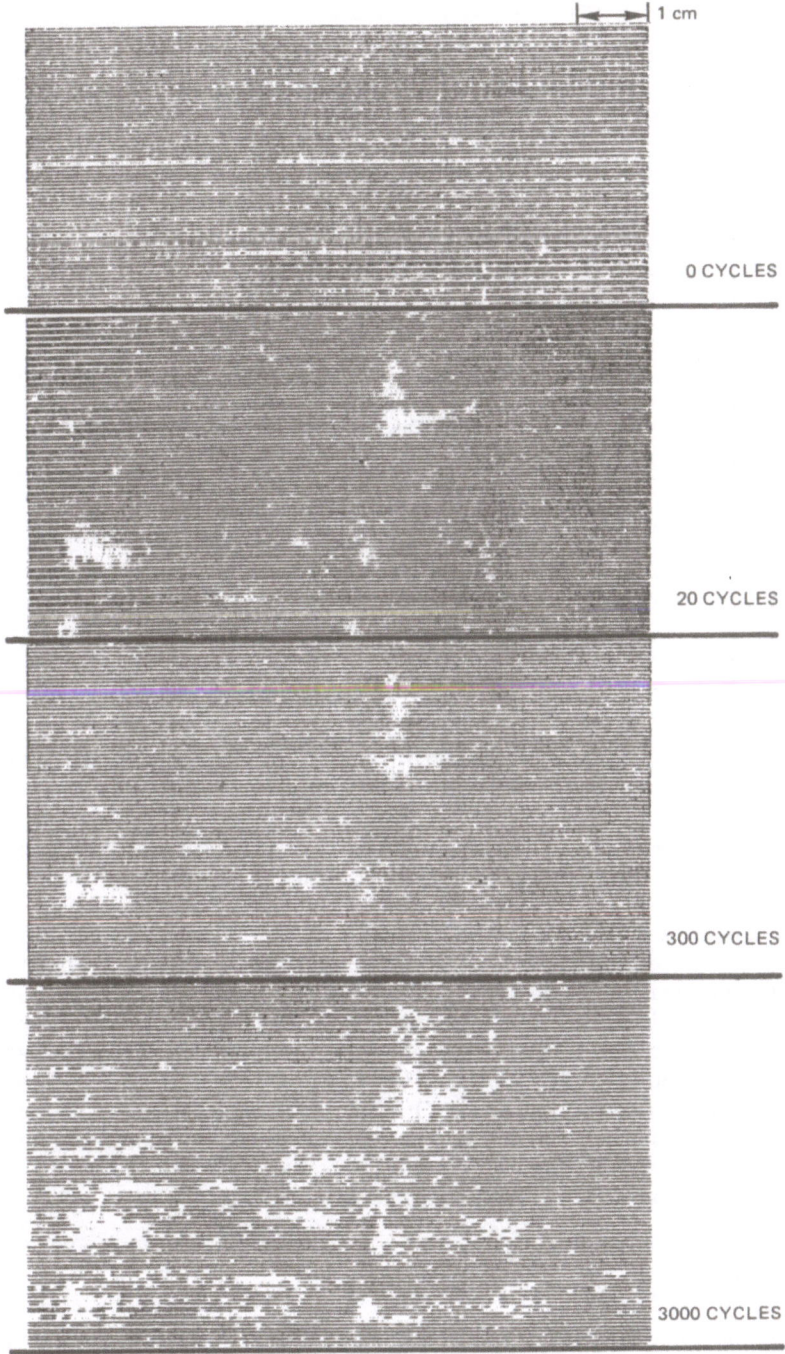

1 cm

0 CYCLES

20 CYCLES

300 CYCLES

3000 CYCLES

Figure 10.   Ultrasonic Transmission-Reflection C-Scan
Data Showing Defect Growth

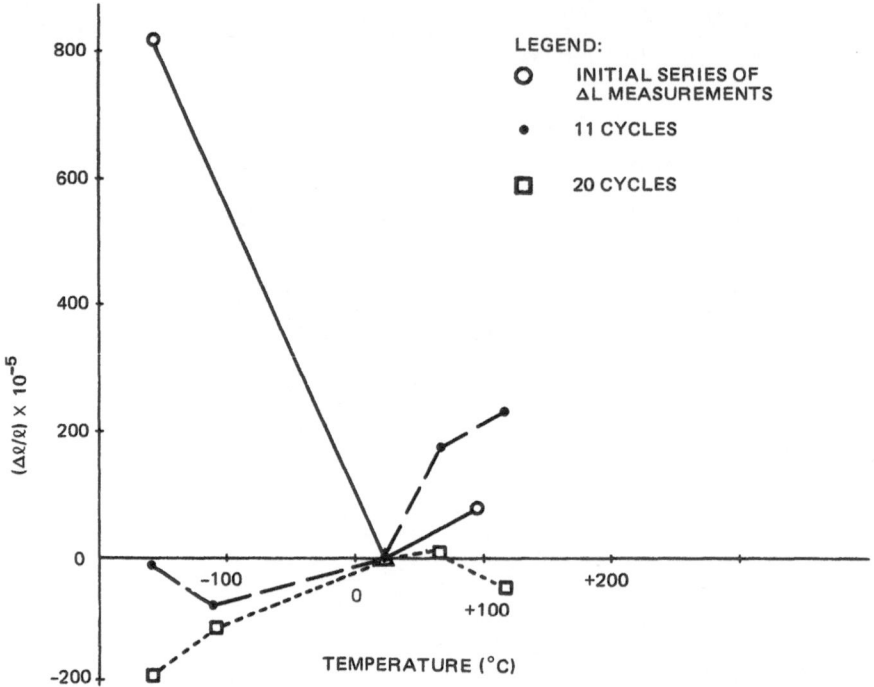

Figure 11.   Expansivity Data for a Graphite-Epoxy Sample Undergoing
Macrocracking During Measurements at Temperature

-160°C) in materials in which measurable defects were not present
(detected) following initial conditioning, ie, up to 15 cycles.

   Optimum test procedures to measure the thermal stability of
graphite-epoxy composites for space applications must consider
practical techniques for qualification of both mechanical and
thermal properties. These considerations include development of
means to minimize measurement error as well as facilitate standard
references through which sensitivity requirements may be determined
for both information (threshold) and repeatability. CTE measurements
should be made in conjunction with complementary nondestructive
testing (ultrasonic C-scan, acoustic emission, etc) to provide
qualitative as well as quantitative prediction data.

                              CONCLUSIONS

   Expansivity characterization of graphite-reinforced composites
is complicated by the anisotropic behavior of the material coupled
with the direct impact of sample thermal history. Neither the
degree nor the mechanism of stress relief is predictable on the

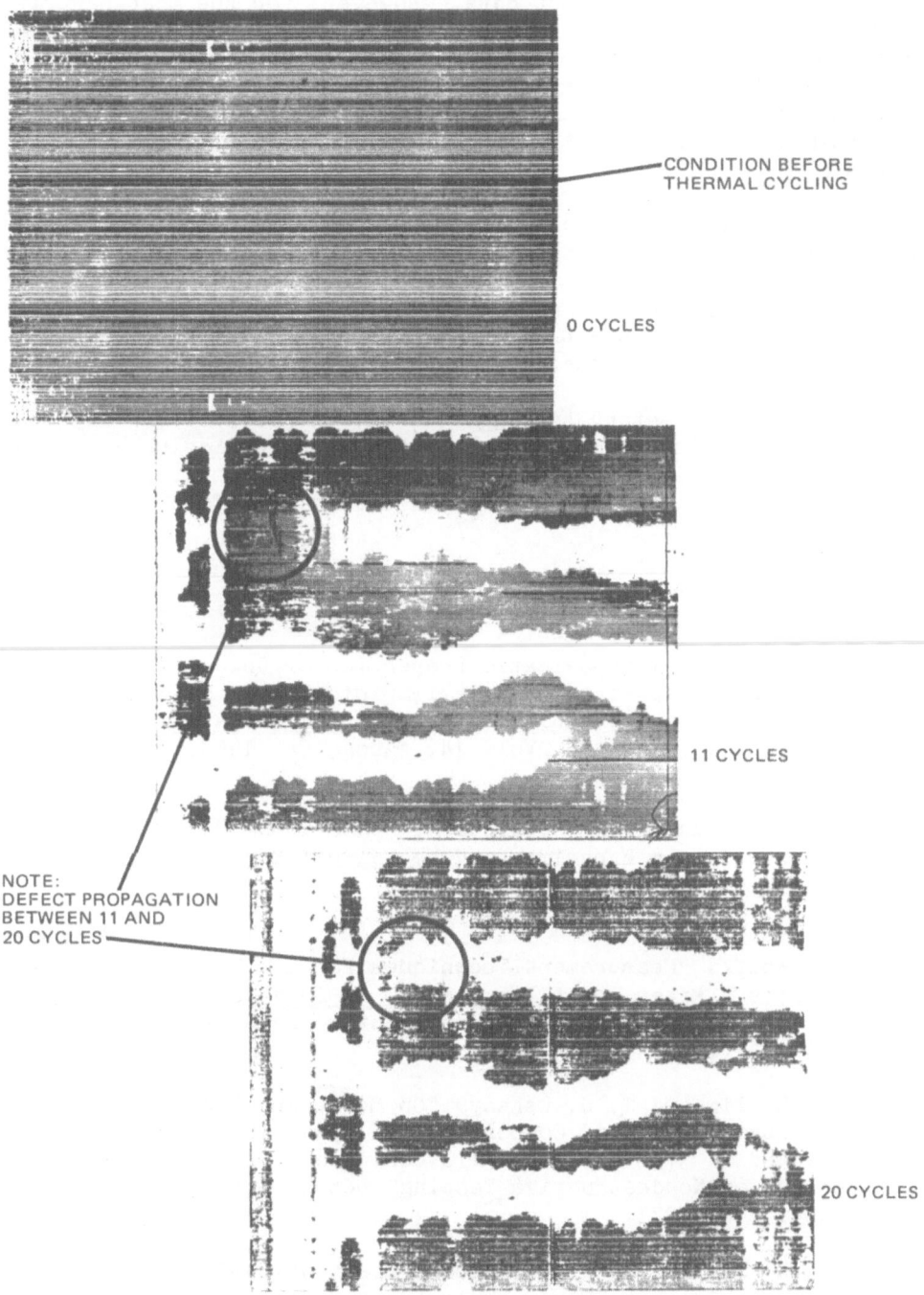

Figure 12. Ultrasonic C-Scan Data Showing Macrochanges

basis of limited (early) test data. However, for the most part it was found that the primary stress relief mechanisms are established during early temperture cycling. Forecasting the eventual capability of understanding and monitoring micromechanisms related to stress, the use of nondestructive testing (NDT) to indicate the probability and mode of stabilization in advanced composites is feasible.

As a complementary tool, NDT techniques have been found to be invaluable for understanding strain relief mechanisms that directly impact sample expansivity. For example, erratic CTE measurements (slope reversal) have been attributed to macro relief mechanisms which result in the nucleation and/or propagation of defects (cracks, disbonds, etc) during thermal cycling.

Use of complementary testing (eg, ultrasonics) to qualify sample stability is recommended. Further study employing acoustic-emission analysis is suggested to provide in situ monitoring of micromechanisms which may link ordered stress relief to stabilization of CTE.

REFERENCES

[1] C. C. Chamis, "Thermoelastic Properties of Unidirectional Filamentary Composites by a Semiempirical Micromechanics Theory," Advanced Techniques for Material Investigation and Fabrication, SAMPE, Vol. 14, Azusa, Ca, 1968.

[2] S. J. Kline and F. A. McClintock, "Describing Uncertainties in Single-Sample Experiments," Mechanical Engineering, January 1953.

[3] E. G. Wolff, "Measurement Techniques for Low Expansion Materials," Materials & Processes - In Service Performance, Vol. 9, National SAMPE Technical Conference Series, October 1977.

[4] L. A. Haslim and T. J. DeLacy, "On Measuring Thermal Stability of Advanced Composites for Space Applications," presented at 37th National Fall Conference, American Society for Nondestructive Testing, Detroit, Michigan, October 3-6, 1977.

[5] I. M. Daniel and T. Liber, "Measurement of Lamination Residual Strains in Graphite Fiber Laminates," Proc. Second International Conference on Mechanical Behavior of Materials, American Society of Metals, Metals Park, Ohio, 1976.

[6]   C. K. H. Dharan, "Thermal Fatigue of Metallized Graphite-
      Fiber Epoxy Composites," AIME-TMS Fall Meeting, Chicago,
      October 25, 1977 (to be presented).

Other Source Material

[7]   R. C. Bruner, "Material Selection for the Space Shuttle
      Orbiter Payload Bay Doors," Bicentennial of Materials Progress,
      SAMPE, Vol. 21, Azusa, Ca, 1976.

[8]   R. A. Stonier and H. L. Hillesland, "Fabrication of a Graphite/
      Epoxy Antenna for the Viking Orbiter Spacecraft," Proc.
      19th Annual SAMPE Symposium, April 1974.

[9]   H. Haydostian, S. A. Robinson, F. A. Taormina, and G. D.
      Walker, "Lightweight Composite Spacecraft Antenna," Proc.
      21st National SAMPE Symposium, April 1976.

THERMAL EXPANSION AND MECHANICAL BEHAVIOR OF GRAPHITE-FIBER-REINFORCED TANTALUM-CARBIDE COMPOSITES

Julius Jortner

McDonnell Douglas Astronautics Company

Huntington Beach, California

ABSTRACT

Thermal expansion measurements, on a variety of graphite-fiber-reinforced tantalum-carbide composites fabricated at temperatures above 3,000°C, imply that little bonding exists between fibers and matrix at room temperature. The results of flexure tests tend to confirm this hypothesis. Heating to above 1,600°C appears to provide significant mechanical interactions between fibers and matrix, presumably caused by differential thermal expansions and interface friction, which lead to highly anisotropic thermal strains at higher temperatures despite isotropic thermal expansion at lower temperatures. One implication is that the thermal expansion behavior of these composites will be similar, in terms of volumetric thermal strain, to that of the carbide matrix, as long as the carbide is the continuous phase of the composite. It was only in those composites which showed evidence of a damaged and/or discontinuous matrix phase that significantly lower volumetric thermal expansions were observed.

INTRODUCTION

Composites of tantalum carbide reinforced with graphite fibers have been made to explore their applicability in high-temperature environments. The high-modulus, high-strength, graphite filaments were intended to reinforce the brittle matrix to confer some thermal stress resistance while improving the strength and stiffness at elevated temperatures.

This paper reviews the thermal expansion behavior to 2,760°C and flexural behavior at 20°C of a number of such composites. The data show several interesting phenomena including changing anisotropy as the composites are heated, large hystereses in thermal strain, and a lack of mechanical reinforcement of the matrix at room temperature. An attempt is made to explain the observations qualitatively in terms of lack of bonding and differences in thermal expansion between matrix and fibers.

MATERIALS

The sample composites discussed here were made with Thornel 50 (T-30) graphite yarn (manufactured by the Union Carbide Corporation) as reinforcement in a matrix made from tantalum carbide powder. Their fabrication included hot-pressing (HP) to consolidate the carbide powder at temperatures in the vicinity of 3,000°C and pressures of 20 MPa. A brief description of each composite, identified by the billet "HP" number, is given in Table 1.

Each billet, typically about 7 cm in height and lateral dimensions, represented somewhat different constructional and process parameters. A variety of reinforcement schemes were attempted (Figure 1) including chopped fibers, chopped yarns, and continuous yarns in unidirectional (1D), bi-directional (2-D), and three-directional (3D) orthogonal arrays. In the chopped-fiber composites, the T-50 fibers were typically less than one millimeter long. The chopped yarn segments were longer, as noted for each chopped-yarn billet in Table 1. Most of the composites were made with yarns first coated with TaC by chemical vapor deposition (CVD). The matrix of some materials included tungsten carbide in relatively small amounts. In some cases, the matrix added to the yarns was a mixture of carbide powders and chopped fibers. In addition to the T-50 reinforcement, some of the materials included carbon from a furfuryl alcohol resin (Varcum) which was used as a binder for the yarns or as an additive to the powdered carbide. The carbide powders were of submicron size to promote infiltration between T-50 filaments.

Composite bulk densities ranged from about 6 to 10 g/cm^3, reflecting a range in carbide content from about 80 to 92% by weight. The balance of the composition was carbon, primarily in the form of Thornel 50 graphite fiber or yarn. In volumetric terms, the carbide content ranged from about 40 to about 60%. Porosities were estimated to be less than 5% in most cases; however, estimates ranged up to 13% for some composites. Non-uniformities in composition and/or consolidation were apparent in many cases. This is reflected in the range of bulk densities

Table 1

DESCRIPTION OF TaC/GRAPHITE MATERIALS

Material Designation (HP No.)	Nominal Weight Percent [1]		Bulk Density (g/cm³)		Measured Weight Percent [3]		Approximate Volumetric Composition, (%) [4]			Distribution of Carbon					Maximum Process Temp (°C)	Remarks [7]
	C	WC	Billet	Specimens [2]	C	TaC+WC	Carbide	Total Carbon	Porosity [5]	Chopped T-50 Filler	Continuous T-50 Yarn 1-D	2-D	3-D	Varcum [6]		
17	11.5	4.8	8.61	8.54–8.60	11.4	88.6	52	48	<1	48					3,220	No CVD TaC
18	14.3	5.3	8.00	8.16–8.25			46	54	<1	54					3,250	No CVD TaC
19	13.9	4.4	7.88	7.61–8.04			47	53	<1	53					3,230	No CVD TaC
38	11	0	8.66	8.41–8.87	11.6	88.4	52	48	<1	48					3,140	No CVD TaC
39	9	0	9.52	9.44–9.64	8.5	91.5	59	39	2	39					3,110	
40	17.8	0	6.63	6.31–6.80			37	57	6	57					3,150	
42	16.4	0	7.01	7.24–7.27	16.3	83.7	40	54	6	54					3,100	Yarn lengths ~10 mm
48	12	17	8.87	8.91–8.97	10.1	89.9	55	43	2						3,000	
49	10	9	9.53	9.46–9.78	7.7	92.3	60	35	5	20					3,000	
50	12.2	0	8.39	8.45–8.56	12.5	87.5	50	50	<1	22					3,100	
51	12.4	9	7.29	7.03–7.85	15.4	84.6	37	50	13	42				8	2,780	No CVD TaC/Varcum in yarns & matrix; yarn lengths ~3 mm
52	11.4	4.7	7.62	7.44–7.70			45	42	13	45					3,010	No CVD TaC
58	12.2	0	8.00	8.68–8.70			48	47	5		21				3,140	No expansion data in Z-direction
88	17.6	5.1	6.67	6.49–6.76	18.5	81.5	37	58	5	15				16	3,100	CVD TaC on yarns prior to Varcum addition; yarn lengths ~6 mm
89	14.6	0	7.46	7.23–7.69	14.7	85.3	44	52	4				32		3,200	Contains NbC. See Reference 1. No CVD TaC. Yarn lengths 6–20 mm. Volume distribution of carbon unavailable.
159[1][5]	~30	0	4.1				27	59	14	✓				✓	3,100	

NOTES: (1) Balance to nominal composition is TaC, except for HP 159 which is 50–50 (by weight) mixture of NbC and TaC.

(2) Range of specimen densities from mechanical test specimen blanks; all densities are mass divided by external volume.

(3) Measured composition by ash analysis.

(4) Volume percents estimated from composition and porosity using 2.1 g/cm³ for carbon and 14.6 g/cm³ for the carbide matrix (except in the case of HP 159 where 10.2 g/cm³ was used).

(5) Porosity estimated by comparing measured billet density to the theoretical density based on composition; where available, the measured carbide content was used instead of the nominal content.

(6) Varcum is a furfuryl alcohol resin which yields a carbon residue on heating.

(7) Except as noted, the T-50 fibers were coated with TaC, usually 20 W/O but varying between 5 and 30 W/O, by chemical vapor deposition (CVD).

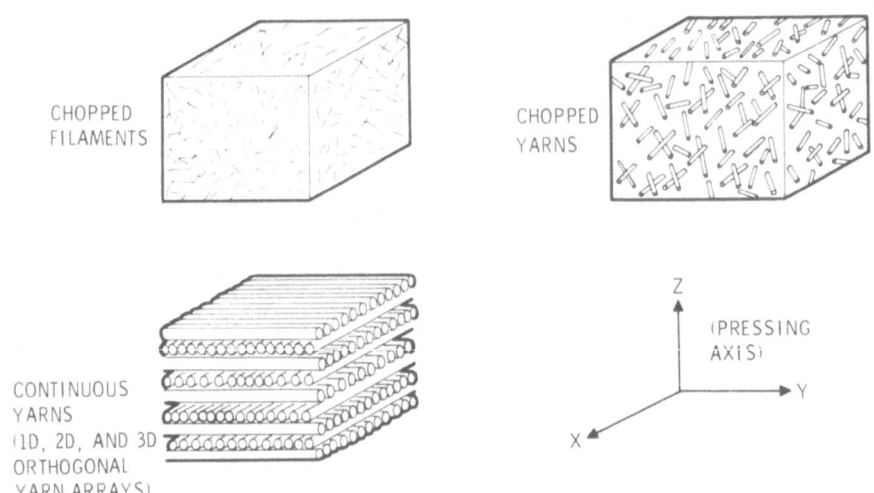

Figure 1. Graphite-Reinforced Metal Carbides

measured on specimens from a single billet and in the differences
that occur between the nominal composition and the carbide content measured by ash analyses on samples of some billets.   The
volume percentages shown in Table 1 are only approximate because
of these non-uniformities and because of uncertainties in the in situ
density of the T-50 filaments and in the fraction of furfuryl alcohol
converted to carbon.

   In addition to the composites made with matrices primarily of
TaC, Table 1 lists HP 159 which was made earlier (Reference 1)
with a 50-50 weight-% mixture of NbC and TaC. Some discussion
of HP 159 is included in this paper because the material exhibited
unusually low thermal expansions.

   Table 2 presents some properties of the carbides and the
Thornel-50 fiber for reference in the discussion of the thermal
expansions and mechanical behavior of the composites.

   The hot-pressing operation, included in the fabrication of all
the composites, preferentially orients and/or distorts the various
constituents, giving a texture that generally results in anisotropy
of properties. In the case of composites made with continuous
yarns, the initial orientation of the yarns also influences the anisotropy of properties. The nomenclature adopted to describe
directionality in the materials is shown in Figure 1: "Z" refers
to the direction of hot pressing; "X" and "Y" are two mutually
perpendicular directions in the plane perpendicular to the Z axis.

Table 2

REPRESENTATIVE PROPERTIES OF CARBIDES AND FIBERS
USED IN FABRICATION OF THE COMPOSITES

	NbC	TaC	WC	Thornel - 50 Fibers
Specific gravity	7.8(a)	14.6(a)	15.6(a)	1.66(g)
Melting pt, °C	3,900(a)	3,880(a)	2,870(a)	-
Young's modulus, GPa at 20°C	330(b)	280(b)	680(b)	360 axial(d) 20 radial(e)
Mean coefficient thermal expansion, $10^{-6}/°C$, 20° - 2,800°C	7-8(c)	7.2-8.6(c)	6(c)	<1.0 axial(d) ~16 radial(f)

Notes regarding sources of data:

a)   Hdbk Chem Phys (44th edition)

b)   From J.H. Westbrook and E.R Stover, Table 9-2, in High Temperature Materials and Technology, edited by I.E. Campbell and E.M. Sherwood, John Wiley & Sons, 1967.

c)   Extrapolations of data presented in Figure 9-8, Ibid

d)   C.R. Rowe and D. L. Lowe, Extended Abstracts of the 12th Biennial Conference on Carbon, Irvine, California 1977.

e)   No data, estimates only.

f)   Based on extrapolation of results of R. C. Fanning and J.N. Fleck, Extended Abstracts of the 10th Biennial Conference on Carbon, Bethlehem, Pa. 1971.

g)   It is likely that the density is increased towards 2.26 g/cm^3, the theoretical value for graphite, during hot pressing. Other properties may also change during processing.

In the case of materials made with chopped filaments or yarns, the X-Y plane is a plane of isotropy and the selection of the X and Y directions is arbitrary. In the case of composites reinforced in more than one direction with continuous yarns, the X and Y directions are selected parallel to yarn directions. In the case of the unidirectionally reinforced composite (HP50), X is taken parallel to the yarns.

EXPERIMENTAL TECHNIQUES

Thermal expansions were measured in inert gas using a quartz dilatometer to 800°C and a graphite dilatometer up to 2,760°C. Heating and cooling rates were in the vicinity of 5 to 10°C per min. The specimen was a bar about 70 mm long and 5 mm square in cross-section which, when the billet was of insufficient length, was built up of two or more pieces in series.

Flexure tests were performed at room temperature at a deflection rate of about 1 mm/min on specimens approximately 5 by 5 mm in cross section by 38 to 50 mm long, using a three-point bending fixture. Stresses were calculated from the applied load and strains were estimated from the measured specimen deflections using the classic linear-elastic beam equations (neglecting shear deformations and assuming the neutral axis is at the center of the cross section). The deflection of the specimen at its midpoint was inferred from the cross-head travel measured at a point as close to the specimen as practically possible. To account for the deflection of the load train an aluminum specimen was tested to provide a standard of known stiffness from which a load-dependent correction factor was calculated and applied in the calculations of strains and stiffnesses. All the specimens appeared to fail in tension. The strength and failure strain information may, therefore, be taken as a first approximation to the tensile behavior of the materials.

For thermal expansion data usually one specimen was tested in each of the material's principal directions. In flexure, duplicate specimens usually were tested. Where more than one specimen was tested, the data reported here are averages.

RESULTS AND DISCUSSION

Thermal Expansion

The thermal strains of each material to 1,100 and to 2,760°C are shown graphically in Figure 2. The following inferences may be derived:

A. In virtually every case the expansion in the Z direction is larger than in the X direction.

B. Anisotropy of thermal strain increases as temperature increases. Expansion is nearly isotropic up to 1,100°C, whereas at 2,760°C the materials exhibit varying degrees of anisotropy.

C. The volumetric expansion of these composites estimated by summation of the linear strains ($e_z + 2e_x$), is nearly equal to that of pure TaC. The TaC data "bars" shown in Figure 2 are based on the information gathered from the literature, Table 2.

D. The chopped yarn material, HP 51, exhibits exceptional behavior in that significant anisotropy is shown at 1,100°C and the volumetric expansion is distinctly less than that of TaC.

E. None of the materials investigated duplicated the exceptionally low thermal expansions reported for HP 159 in Reference 1. HP 159 has a matrix of NbC and TaC reinforced with chopped Thornel 50 yarn. As NbC has

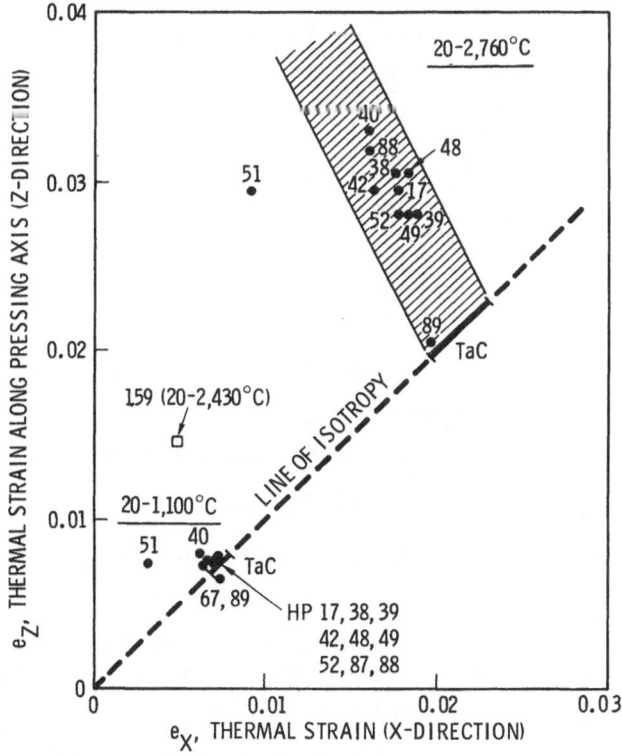

Figure 2. Thermal Expansion of TaC/Graphite Composites to 1,100°C and 2,760°C; Shaded Region Between Lines Has Same Volumetric Strain ($e_z + 2e_x$) as Range of TaC Data

nearly the same thermal expansion as TaC, the reasons
for the markedly lower expansion of HP 159 are not
immediately clear.

Figure 3 shows that the permanent sets observed at room
temperature, after the thermal expansion specimens were cycled
once to 2,760°C, are always positive (permanent growth) in the
Z direction and usually negative (permanent shrinkage) in the
X direction. HP51 is the only material in Figure 3 which grew in
the X-direction as well as in the pressing direction. HP 159 also
grew in both directions (Reference 1). Also plotted in Figure 3
is the line representing zero net volume change after temperature cycling. While some materials come quite close to the line,
most exhibit some permanent expansion after cycling. An exception is HP 89, the only three-directionally reinforced billet for
which data are available, which exhibited net shrinkage.

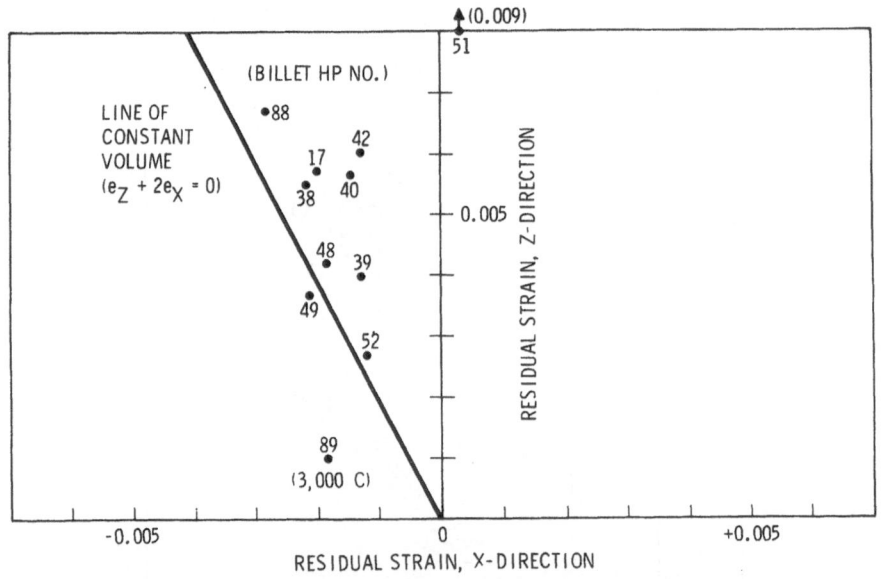

Figure 3. Permanent Length Changes at 20°C After Heating to 2,760°C

Further insight into the behavior of the TaC/graphite composites may be gained from the curves shown in Figures 4 through
7. The thermal strain response typical of the chopped-filament
materials is shown in Figure 4. The features already alluded to
are readily seen: the near isotropy to about 1,000°C and the
permanent sets on cooling. Since the Thornel yarn is highly
anisotropic in its expansion at all temperatures, the behavior
shown in Figure 4 suggests that the fibers do not exert any rein-

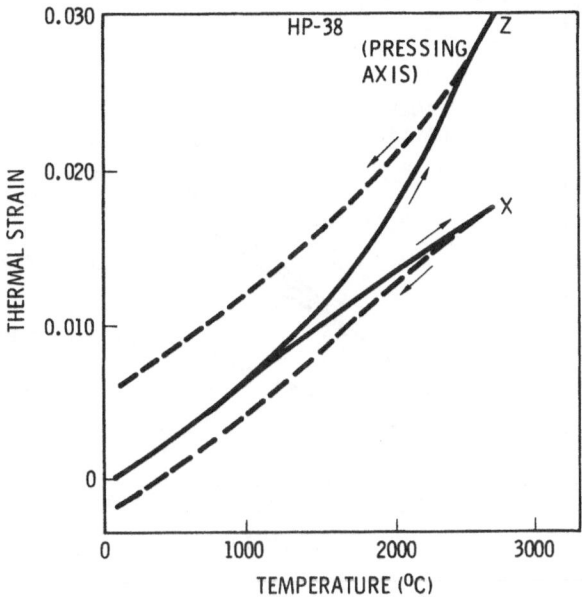

Figure 4. Thermal Expansion of HP-38, a Chopped-Filament Composite

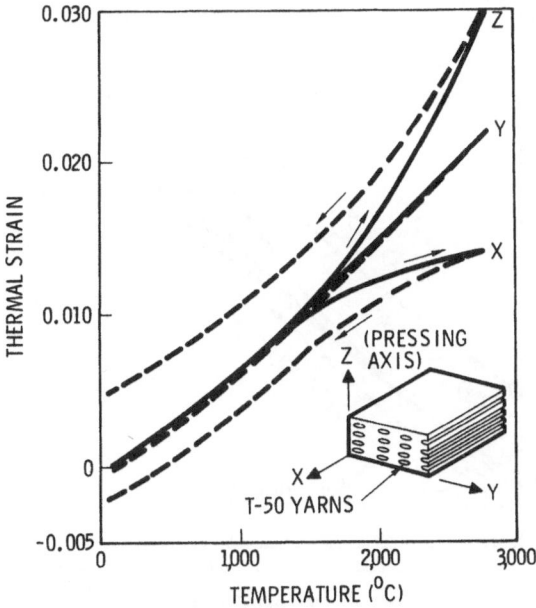

Figure 5. Thermal Expansion of HP-50, a Unidirectionally Reinforced Composite

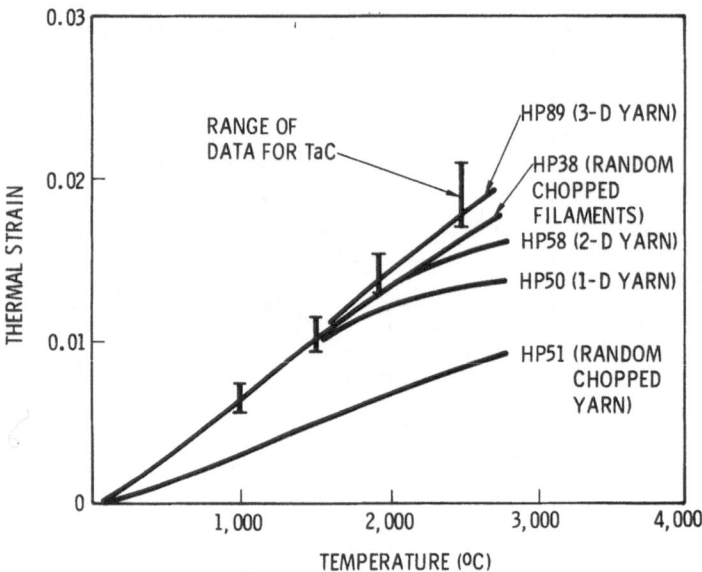

Figure 6. Thermal Expansion in X-Direction For Different Degrees of Reinforcement Alignment

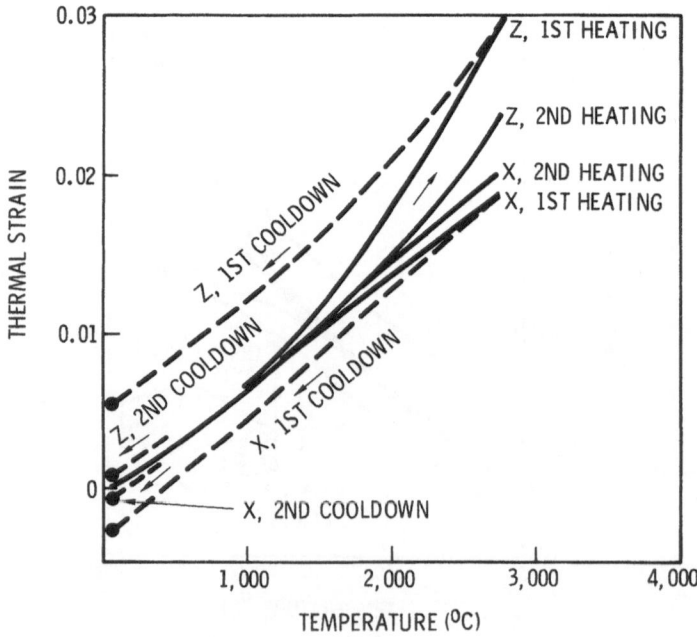

Figure 7. Thermal Expansion of HP-17 Upon Repeated Heating of Each Sample

forcing effect on the composite below about 1,000°C; in other words, the filaments are not effectively bonded to the TaC matrix. This appears plausible if one considers the filaments and matrix to have been in intimate contact at the maximum processing temperature (above 3,000°C). The large shrinkage of the graphite filaments transverse to their axes upon cooling to room temperature probably results in a debonding between filaments and matrix. Then, if the matrix of TaC is the continuous phase in the composite, the composite thermal expansion will be essentially that of TaC until the filaments have expanded sufficiently to again make contact with the TaC. Above the effective contact temperature, the expansion of the composite transverse to the preferred filament orientation (i.e., in the Z direction) would be increased over TaC alone. The expansion in the direction of the filaments would be decreased by one or both of these mechanisms: frictional forces at the interface between filament and TaC matrix may restrain the TaC expansion and/or, the creep of the TaC normal to the filaments may cause Poisson's shrinkage of the TaC in the direction of the filaments.

The results shown in Figure 5 for HP 50, the unidirectionally reinforced continuous-yarn composite, also show isotropy up to more than 1,000°C. As indicated by the sketch in Figure 5, there are three distinct principal directions in the HP 50 composite. Each direction exhibits different behavior above 1,200°C. The behavior in the Z direction (the pressing direction) is virtually equal numerically to the behavior of HP 38 (Figure 4) in its pressing direction. The fact that there is significant difference between the Y and Z directions in HP 50 is perhaps due to the flattened geometry of the yarns which probably reduces the mechanical effectiveness of the graphite (in forcing the TaC to expand) in the Y direction as compared to the Z direction.

The effectiveness of the Thornel 50 reinforcement in reducing the expansion of the composites is illustrated in Figure 6. Increasing collimation of the filaments, as represented by the sequence

HP 89, three-directional orthogonal array of yarns,
HP 38, random (but pressed) array of cut filaments,
HP 58, two-directional laminate of yarns,
HP 50, one-directional yarns,

tends to reduce the thermal expansion in the X-direction at temperatures above 1,600°C. A distinctly different behavior is exhibited by HP 51 which was reinforced with a random array of chopped yarns that had been impregnated with furfuryl alcohol (Varcum) prior to mixing with the TaC. The Varcum resin converts to carbon during heating; organic gases are given off during its pyrolysis. It appears probable that the Varcum impregnation of the yarns, without first coating the yarns with TaC by chemical vapor deposition (CVD), is responsible for the low thermal expansion of HP 51. Several other chopped-yarn composites (HP 42, 49, and 88) were also tested for thermal expansion but, as Figure 2 shows, their thermal expansions were not significantly lower than the other types TaC/graphite composites. HP 42, 49, and 88 were all made with yarns that had first been CVD coated with TaC. The only composite other than HP 51 that had been made with Varcum-impregnated yarn without CVD TaC was HP 159, which also exhibited exceptionally low thermal expansions.

Understanding of the mechanisms by which the thermal expansion is lowered is incomplete. One may speculate that the Varcum causes the yarn to act as a larger, rigid unit of graphite which is a more effective reinforcement than a bundle of filaments coated with, but not really bonded to, TaC. This leaves open the question of why load transfer would be more effective between a large graphite unit and TaC; indeed, except for mechanical attachment via crevices in the yarn unit, it is difficult to see how increased size of the graphite unit would improve load transfer. Another hypothesis is that the large graphite unit may, during the heating and cooling steps that occur in processing, exert enough force to crack the TaC matrix so it no longer acts as a continuous body of TaC. This latter idea receives support from the fact that both HP 51 and HP 159 had exceptionally low mechanical strengths. Both materials also had relatively large porosities, perhaps a consequence of the presumed damage to the matrix That porosity by itself is not the controlling factor in thermal expansion may be seen from the data for HP 52, a continuous yarn composite which exhibited thermal expansions similar to most of the composites tested (Figure 2) but had porosity approximately equal to that of HP 51. The fact HP 51 was processed at a lower temperature than the other composites may be influential; HP 159, however, had been processed at the normal high temperature.

The hysteresis and permanent set characteristics of the thermal expansion behavior of these materials probably deserve further attention One set of specimens, from HP 17, was retested after the first measurements to 2,760°C. The results (Figure 7) suggest that the thermal strains on reheat are given, to a good approximation, by the strains shown for the first cooldown curve minus the permanent set. Some implications are illustrated in Figure 8 where the second-cycle strains are estimated from the permanent sets for a number of composites. Second-cycle behavior is expected to be generally more isotropic than first-cycle behavior. Figure 8 also shows that anisotropy of thermal strain at high temperature tends to increase with increased graphite content, an effect related to that illustrated in Figure 6.

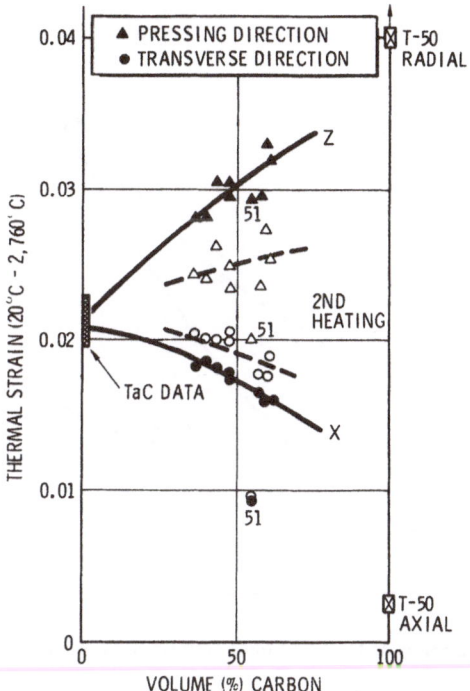

Figure 8. Compositional Effects on Thermal Expansion and Estimated Response After Heating to 2,760°C (Open Symbols)

The reasons for the hysteresis and permanent set are probably related to the poor contact at room temperature between the graphite filaments and the TaC. If the filaments and the surrounding TaC no longer register perfectly because of machining (relief of residual strains, etc) or even just the slight displacements that might occur on cool-down of the billet after hot pressing, then the transverse expansion of the filaments during the thermal expansion test will impose mechanical forces on the TaC which will cause the TaC to creep. The creep strains normal to the predominant filament orientation will tend to be positive. Since creep is a near-constant-volume process, the creep strain in the TaC will be negative in the direction of the filaments, in agreement with the observed results. Repeated thermal cycling should reduce the hysteresis by enlarging the holes in which the filaments reside, and the composite's thermal expansion should approach more closely the isotropic expansion of TaC.

The behavior of HP 51 (Figure 8) is exceptional, indicating that another mechanism is probably at work. The fact that both permanent sets are positive (Figure 9) appears to support the notion of a TaC matrix being cracked by the graphite yarn units.

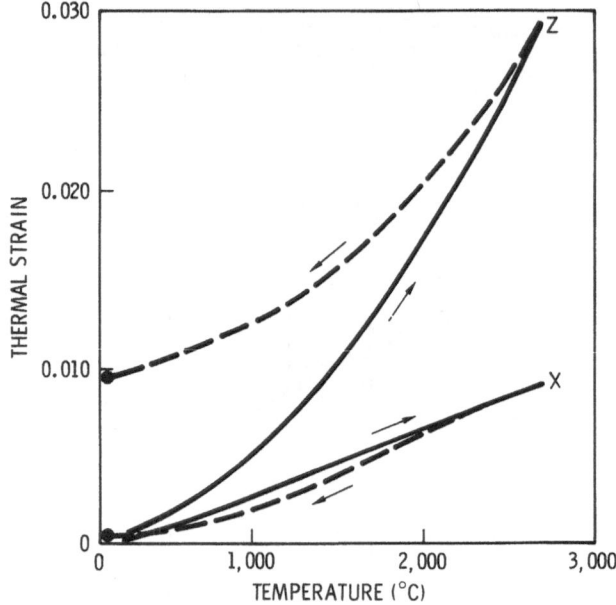

Figure 9. Thermal Expansion of HP-51, Showing Positive Permanent Set in Both Directions

Flexure

Some of the stress-strain curves derived from the flexure tests are shown in Figures 10 and 11. Figure 10 shows two different behaviors: abrupt failure after essentially elastic loading, exemplified by HP 17 in Figure 10; and continuous yielding behavior with no clear failure strain, exemplified by HP 51. The curves shown in Figure 11 for the unidirectional composite, HP 50, reveal the role of yarn continuity in providing some residual load carrying capacity after the initial abrupt failure. (Note that the stresses and strains shown after the load has dropped abruptly in a flexure test may not be accurate because the equations used to calculate them do not account for the existence of partial cracks in the beam; the post-failure stress-strain information is useful only as a qualitative indicator of residual integrity.) The multidirectionally-reinforced continuous-yarn composites usually showed some residual load capacity similar to that of HP 50 in the yarn direction.

Only HP 51 and HP 159 showed continuous yielding behavior. All the other materials were nearly linear in their loading behavior. The strength of these two chopped-yarn materials in the X direction was below 35 MPa, whereas the other chopped-yarn

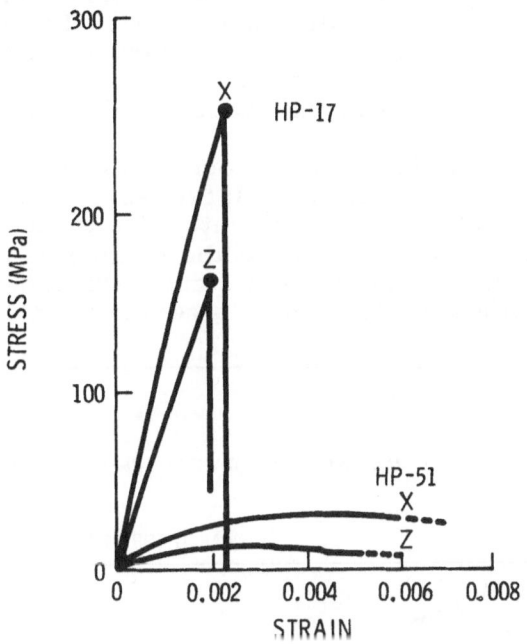

Figure 10. Contrast in Flexural Stress-Strain Curves

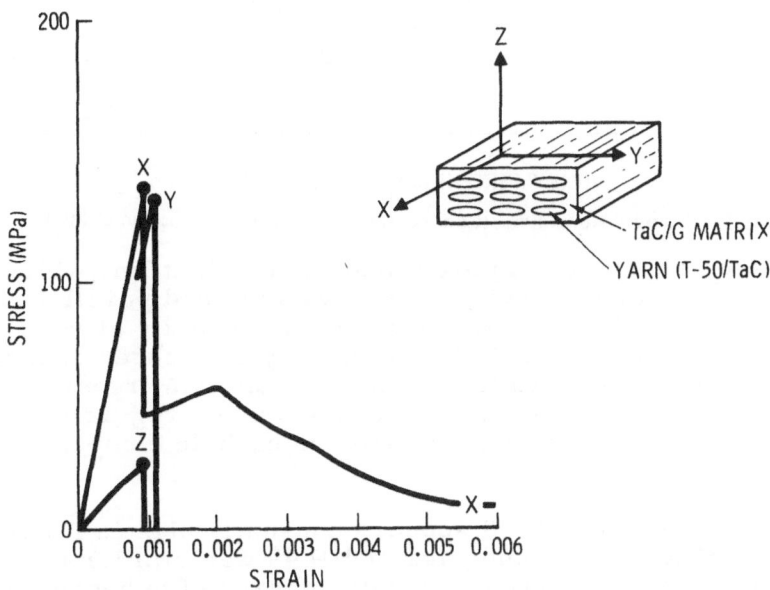

Figure 11. Flexural Strain Responses of 1D Composite, HP-50

materials (HP numbers 42, 49, and 88) had strengths ranging
from 70 to 200 MPa. Only HP 51 and 159 had strains corresponding to maximum stress greater than 0.003. These facts appear
to be related to the exceptionally low thermal expansions of these
two billets by means of the precracked, or otherwise damaged,
TaC matrix hypothesized in the preceding discussion

Compositional influences on Young's modulus in flexure are
shown in Figure 12. Stiffness is decreased, on the average, by
increasing the reinforcement content. The trend of the data is
similar to the behavior expected if the filaments acted as pores.
To illustrate this, two theoretical trends are shown for pore
influence on stiffness: MacKenzie's curve (from Reference 2)

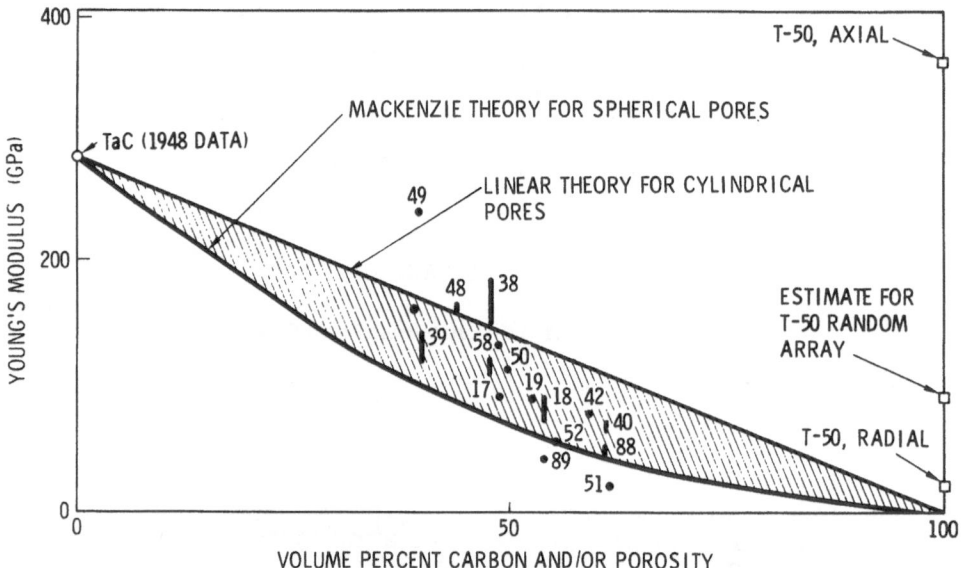

Figure 12. Compositional Effects on Young's Modulus in X-Direction at 20°C

which assumes spherical pores; and the straight line which is
equivalent to assuming cylindrical pores aligned parallel to the
load axis. The idea of graphite acting as porosity at room temperature is consistent with the previously-expressed supposition
that a gap exists between filaments and matrix at room temperature. A similar hypothesis has been advanced regarding the
behavior of graphite-reinforced silicon-carbide composites
(Reference 3).

The stiffness of HP 49 appears unusually high in Figure 12.
The thermal expansion behavior of HP 49 was similar to most of
the composites, suggesting a similar degree of unbonding between
T-50 fibers and the matrix. Thus the stiffness of this billet

appears anomalous. Two possible contributing factors are that (1) some uncertainty attaches to the reference value of Young's modulus for pure TaC, and (2) potential errors in the flexure data itself are greater for the stiffer materials because of the correction noted in describing the flexure technique.

In the direction of pressing, the Z-direction, all the measured Young's moduli fell at or below the MacKenzie curve.

CONCLUSIONS

The behavior of graphite-fiber-reinforced tantalum-carbide composites is illustrative of phenomena that can occur in materials fabricated at very high temperatures. The changes in anisotropy of thermal strain as the composites are heated, and the apparent lack of reinforcement between graphite fibers and carbide matrix at low temperatures, may be tentatively attributed to the large mismatch between free thermal strains of fiber and matrix which causes debonding or cracking on cool-down from the maximum process temperature. The data suggest the continuous matrix phase will control the volumetric thermal expansion of the composite. Lowering the volumetric thermal expansion appears to require a discontinuous matrix. Producing microstructural damage in the matrix during processing may lower the thermal expansivities at the expense of mechanical strength. The suggestion that micromechanical creep processes are involved in thermal expansion at high temperatures probably deserves further study, possibly by conducting thermal expansion tests under conditions of rapid heating.

REFERENCES

1. J. O. Gibson. Low-Linear Thermal-Expansion NbC-Graphite Composite Studies. Phase II Report under Contract SNPC-67 for NASA, McDonnell Douglas Astronautics Company Report MDC G1692, May 1970.

2. W. D. Kingery. Introduction to Ceramics. Chapter 17, J. Wiley & Sons, Inc., 1960.

3. G. W. Hollenberg and R. L. Crane. Effect of Graphite Fiber Additions on the Grain Growth and Strength of Hot-pressed Silicon Carbide. Paper 9-C-74, 76th annual meeting of the American Ceramic Society, May 1974.

ACKNOWLEDGMENTS

The work reported was sponsored by the Defense Nuclear Agency, Washington, DC. The composite materials described were designed and fabricated by J. O. Gibson of the McDonnell Douglas Astronautics Company. The thermal expansion data were measured at the Southern Research Institute, Birmingham, Alabama. The author is indebted to H. S. Starrett of the Southern Research Institute for the observation that the volumetric thermal expansion of most of the composites is the same as that of the carbide matrix. Thanks are due to E. G. Wolff for his constructive review of the first draft of this paper.

THE THERMAL EXPANSION BEHAVIOUR OF FILAMENT WOUND COMPOSITE TUBES

A.A. FAHMY AND C.H. CHIANG

NORTH CAROLINA STATE UNIVERSITY

RALEIGH, NORTH CAROLINA 27607

B.M. HALPIN, JR.

ARMY MATERIALS AND MECHANICS RESEARCH CENTER

WATERTOWN, MASSACHUSETTS 02172

INTRODUCTION AND REVIEW

Fiber reinforced composites have been developed for a large variety of applications and considerable effort has been devoted to their characterization and testing. These materials may actually be "designed" and built to possess any of a wide range of property combinations. The emphasis thus far has been on high strength and stiffness combined with light weight. More recently, some attention has been given to other nonstructural aspects such as electrical and thermal properties, including thermal expansion (1).

Knowledge of the properties of the fiber and matrix, their proportions and configurations makes possible the calculation of the logitudinal and transverse thermal expansion coefficients of unidirectional composites, the former with greater ease and certainty than the latter. Once this is done, it is possible to predict the thermal expansion coefficients of certain balanced composite laminates along two symmetry axes in the laminate plane (2-6). It is also now recognized that the thermal expansion along the thickness direction of these laminates - which is a transverse direction to each layer or ply in the laminate - is quite different from the unidirectional transverse expansion (7,8). However, among the most potentially useful applications of the concept of fiber reinforce-

ment are filament wound constructions such as tubes, and very little
is known about their thermal expansion. A recent examination of a
number of filament wound polymer matrix tubes has shown clearly that
they generally contain a high level of residual stresses (9). This
was made evident by the distorsion the tubes underwent upon slitting.
These stresses have their origin in a number of effects including
the fiber tension during winding, the curing shrinkage of the resin,
and the thermal expansion or contraction of the tube material upon
cooling from the curing temperature. So, while the thermal expansion
behavior of the tubes per se may be of prime interest in some applications, its effects should always be taken into account whenever a
stress analysis of these tubes is attempted.

This paper thus deals with the thermal expansion behavior of
filament wound composite tubes and the thermal stresses induced
therein in the absence of thermal gradients. It represents one further step in a continuing search for a better understanding of the
thermal expansion of fiber composites.

CHARACTERIZATION OF THE TUBE MATERIAL : THEORY AND EXPERIMENT

Calculation of the thermal expansion of filament wound tubes
depends not only on the expansion behavior of the tube wall material
but also on its elastic properties. The tube may be considered to
consist of shells or layers of aligned fibers in a resin matrix with
the angle between the fibers and the tube axis - known as the winding
angle - alternating from $+\theta$ to $-\theta$ in alternate layers. If the number
of layers is large enough the wall material may be treated as angle-
ply laminates which are orthotropic with material axes r, θ and z
of the tube (Fig. 1). The θ and z axes correspond to the in-plane
symmetry axes 1 and 2 while the r axis points in the thickness direction 3 (Fig. 2).

It was necessary to determine the in-plane and the thickness
thermal expansion coefficients of the tube material. It soon also
became evident that there is considerable lack of information on any
elastic property of fiber reinforced laminates involving the thickness direction. So, it was decided to determine - experimentally and
analytically - all elastic properties needed for the calculations,
namely Young's moduli E_z, E_θ and E_r and Poisson's ratios $\nu_{z\theta}$, $\nu_{\theta r}$
and ν_{rz} all of which are independant of one another. To this end we
embarked on the tedious task of laying up relatively thick angle-ply
laminates from Kevlar-Epoxy prepreg tape and autoclave curing them.
The ply angles were 0° (unidirectional), $\pm 15^\circ$, $\pm 30^\circ$, and $\pm 45^\circ$ (cross-
ply). To avoid layup errors and to be able to cut individual samples
out of a large laminate rather than making them individually, the
fiber orientations were made to alternate from $+\theta$ in one ply to $-\theta$
in the next all the way through the thickness. Thus, the laminates

did not possess midplane symmetry, but for a laminate several hundred layers thick this is of practically no consequence. Samples were then cut out, carefully machined and tested.

Thermal expansion coefficients between 50°C and 150°C were determined using an Orton automatic dilatometer at a heating rate of 1°C/min. The first heating and cooling cycle showed some hysteresis, presumably associated with driving off moisture, but the behavior became reversible in subsequent cycles. Data used for the determination were obtained from the second cycle.

Mechanical tests - the details of which will be given elsewhere- were performed using a Tinus-Olsen universal testing machine. The strain was monitored by the use of electrical strain gauges.

The ELAS 75 computer code (10) which is a linear elasticity code capable of treating three dimensional anisotropic solids by finite element analysis was used to evaluate the elastic properties of the laminates based on the experimentally determine properties of the unidirectional material. Plain stress elasticity theory was also applied to determine the in-plane elastic properties of the laminates and yielded identical values to those obtained by computer analysis. Thermal expansion coefficients were also calculated based on the longitudinal and transverse coefficients of the unidirectional material as determined experimentally.

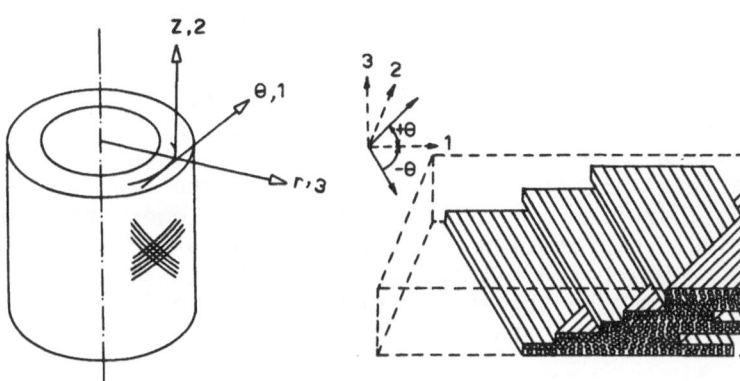

Coordinate Axes For Tube Analysis   Lay up of pre-preg tape in the preparation of Kevlar-Epoxy angle-ply laminates.

Fig. 1   Fig. 2

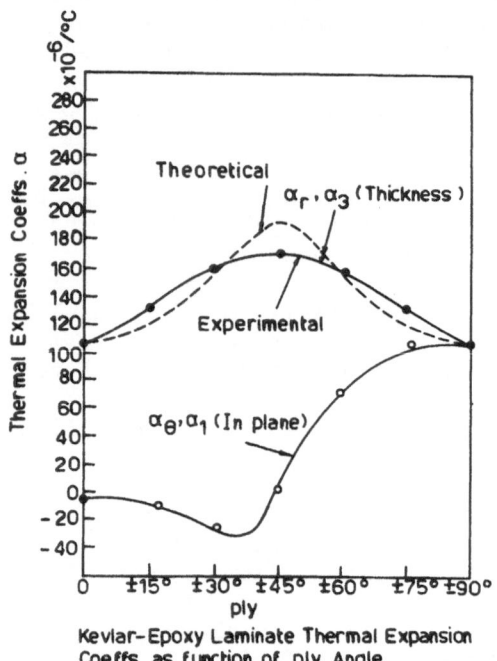

Kevlar-Epoxy Laminate Thermal Expansion
Coeffs. as function of ply Angle.

Fig. 3

Results of the thermal expansion work are given in Figure 3.
It is evident that the coefficient α_θ for the $\pm\theta$ laminate is of course the same as α_z for the $\pm(90^\circ - \theta)$ one, and that α_r for the $\pm\theta$
laminate and the $\pm(90^\circ - \theta)$ laminate are identical. Worthy of note
is the high negative value of the inplane coefficient of laminates
of ply angle around 35° and the very high positive value of the
thickness coefficient of the cross-ply laminate. The fact that the
laminates exhibit in-plane coefficients which are lower than the lowest for the single ply(the longitudinal coefficient), and also thick
ness coefficients which are higher than the ply highest (the transverse coefficient) may be thought of as "Poisson's effects" resulting
from the mutual constraint the layers impose on one another in the
laminate.

Figures 4 and 5 show the experimentally and analytically determined values of the elastic properties of the angle-ply laminates.

ANALYSIS OF THE FILAMENT WOUND TUBES

The next phase of this work was to calculate the change in the
inner and outer radii of filament wound tubes as a result of a unit
temperature change and hence "coefficients" of thermal expansion.

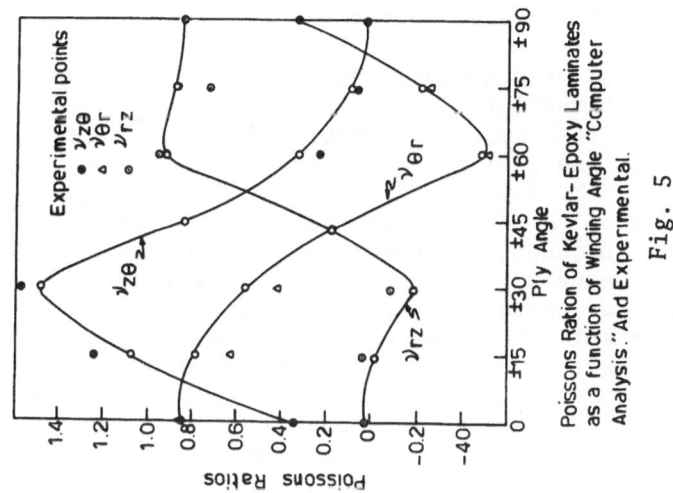

Poissons Ration of Kevlar-Epoxy Laminates as a function of Winding Angle "Computer Analysis" And Experimental.

Fig. 5

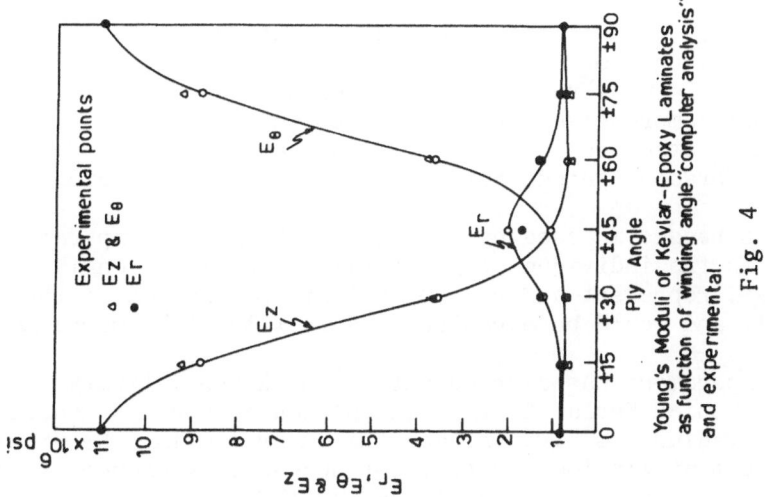

Young's Moduli of Kevlar-Epoxy Laminates as function of winding angle "computer analysis" and experimental.

Fig. 4

This was done by the use of the ELAS 75 computer code. The assumption was made that the tube material is orthotropic having the elastic and thermal expansion properties of the corresponding angle-ply material. The analysis was conducted on a thin sector of the tube 4° wide, divided into 36 hexahedral elements.

Results of this analysis are shown in Figure (6) for tubes of outer to inner radii ratios 2, 1.5 and 1.2. It can be seen that thermal expansion coefficient of inner and outer radii of thin tubes are both close to the inplane coefficient of the corresponding angle-ply laminate i.e. the laminate of the same angle as the fibers make with the hoop direction of the tube. As the wall thickness increases the much larger thickness expansion coefficient starts to play a significant role, causing the inner radius to expand less and the outer radius to expand more than the corresponding angle ply material. In addition the wall thickness expansion coefficient is less than the thickness expansion coefficient of the laminate. It is therefore obvious that there exists a state of stress in the tube wall as a result of temperature change. Computer analysis shows there is a radial compressive stress which is, of course, zero at both the inner and outer surfaces and reaches a maximum at some radius in between and that there are hoop stresses which range from a maximum tension at the outer surface to maximum compression at the inner surface. With the tubes having free ends, the axial stresses were found to be insignificant and the axial thermal expansion coefficient to be essentially the same as that of the corresponding angle ply laminate.

Figure (7), shows the stress distribution in the walls of ±60° Kevlar-Epoxy tubes of different wall thickness. This distribution is typical of all tubes regardless of winding angle. The maximum tensile and maximum compressive stresses are shown in Figure (8), and the maximum radial stresses in Figure (9), all as functions of the winding angle for three different ratios of outer to inner radius ro/ri.

The stresses induced by a moderate change in temperature, when referred to the individual ply material axes, i.e. longitudinal and transverse, represent a far from negligible fraction of the strength of the ply and would be expected to affect the tube performance.

An experiment was carried out to check the validity of the analysis. One thick Kevlar-Epoxy filament wound tube was manufactured. The inner radius was 3.25 in. and the outer 4.0 in. The winding angle at the start of winding was 45°, but necessarily changed slightly during winding reaching a value of 49° on the outer surface, giving a winding angle of 47° at midradius. A ring cut out of this tube was instrumented with electrical strain gauges on the inner and outer surfaces to measure hoop strains, and a compensating gauge was cemented to a flat fused silica plate. Both the ring and plate were placed in an electrically heated oven and strain readings taken during the se-

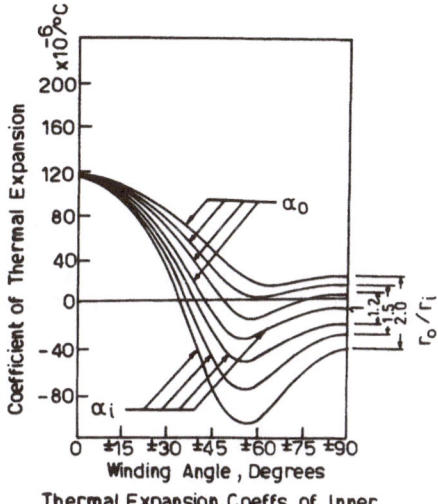

Thermal Expansion Coeffs. of Inner and Outer radii as Function of Winding Angle .

Fig. 6

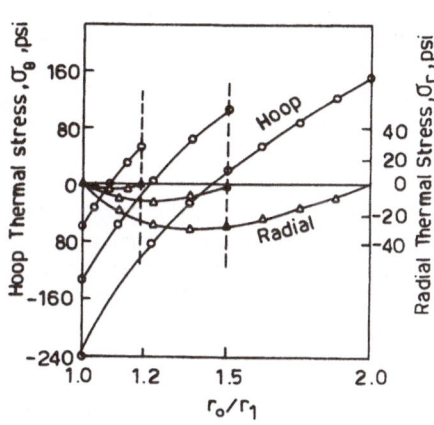

Hoop and radial thermal stress distribution, for ΔT=1°C, through thickness of the +/-60° Kevlar-Epoxy tubes with free ends and r_o/r_1 =1.2 ,1.5 and 2.0.

Fig. 7

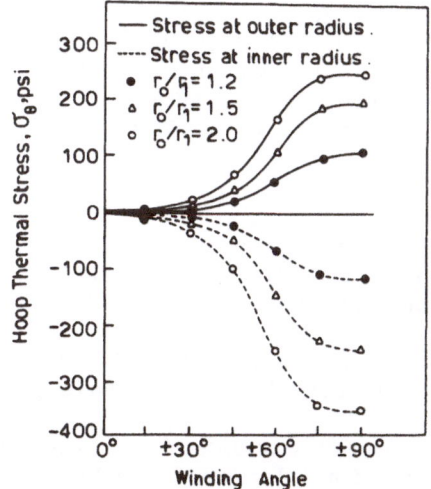

Hoop thermal streses in the inner and outer layers due to ΔT=1°C, as function of winding angle , of the Kevlar – Epoxy tubes with free ends and r_o/r_1 = 1.2 1.5 and 2.0.

Fig. 8

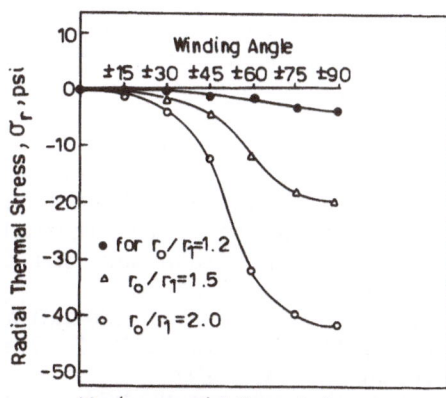

Maximum radial thermal stresses as function of winding angle, due to 1 °C increase in temperature , along individual thicknesses of Kevlar-Epoxy tubes with free ends and r_o/r_1 = 1.2 ,1.5 and 2.0 .

Fig. 9

cond heating and cooling cycle. The expansion coefficient α_θ of the inner and outer radii were found to be -26.5 x 10^{-6}/$^\circ$C and +17.5 x 10^{-6}/$^\circ$C. Analytical values were -24.2 x 10^{-6}/$^\circ$C and +15.4 x 10^{-6}/$^\circ$C for a winding angle of 47°. This agreement is remarkable considering the experimental difficulties, especially those involved in making composites of uniform quality, and confirms the validity of the approach.

In conclusion this work underlines the fact that by the proper choice of the winding angle and the ratio of outer to inner radius it is possible to exercise considerable control over the thermal expansion behavior of filament wound tubes and the thermal stresses induced therein.

ACKNOWLEDGEMENTS

The authors wish to express their appreciation to Mr. Stanly Tozlowski for his assistance in sample preparation and Mrs. Nadia ElMasry for her help in conducting the thick tube test.

REFERENCES

1. Hashin, Z. Designing for Non-structural Applications and Physical Properties. Rapporteur's report, Proceedings of the 1975 International Conference on Composite Materials, Vol. 2, 472.
2. Ashton, J.E., Halpin, J.C. and Petit, P.H. Primer on Composite Materials. Analysis (Technomatic, Stamfrod, Conn., 1969).
3. Halpin, J.C. and Pagano, N.J. Technical Report AFML-TR-68-395, 1969.
4. Fahmy, A.A. and A.N. Ragai. Thermal Expansion of Graphite-Epoxy Composites. Journal of Applied Physics, Vol. 41, p. 5112, 1970.
5. Fahmy, A.A. and A.N. Ragai ElLozy. A Discrete Element Method for the Calculation of the Thermal Expansion Coefficient of Unidirectional Fiber Composites. AIP Conference Proceedings, No. 17, Thermal Expansion, 1973.
6. ElLozy, A.R. Thermal Expansion Behavior of Particle and Fiber Composite Materials. Ph.D. Thesis, N.C. State University Raleigh, North Carolina 1973.
7. Fahmy, A.A. and A.N. Ragai ElLozy. Thermal Expansion of Laminated Fiber Composites in the Thickness Direction. J. Composite Materials Vol. 8, p. 90, January 1974.
8. Pagano, N.J. Thickness Expansion Coefficient of Composite Laminates. J. Comp. Mats., page 310-12, July 1974.
9. Fahmy, A. Examination of Filament Wound Tubes. Interim Progress Report, submitted to AMMRC (batelle Task No. 75-333) 1975.
10. Utku, S. Elas' 75 Computer Program for Linear Equilibrium Problems of Structures. Computer Software Management and Information Center, University of Georgia, Athens, Georgia, December 1971.

APPENDIXES

ENERGY STORAGE AND THE CRUISING SAILBOAT - BANQUET SPEECH

Dr. P.J. Dyne

Director, Office of Energy Research & Development
Energy, Mines and Resources
Ottawa, Ontario

One of the objects of a good holiday is to leave one's work behind. Deciding to cruise in a little sailboat this July in the Thousand Islands region of the St. Lawrence turned out to be an excellent choice. While I've sailed dinghies a fair amount, this was the first time that I sailed a bigger boat. It was really great fun - one was indeed isolated from the dreary worries of today's world; no telephone, no newspaper; the radio only for weather forecasts.

Although one is isolated, one is hardly roughing it. It seems to be the object of the game to bring as many of the comforts of home and cram them into an 8 x 10' space with 5' headroom for four bunks, storage, cooking, eating, dressing, washing and going to the can! It's as if one compressed as many of the comforts of home into a little bubble and gently bounced this bubble off the outside world of wind, water, wave and rock - to have somehow the best of both worlds.

Of course, one never really leaves one's work behind. Because of my interest in Energy, I've got very interested in where, how and why we need energy. I found that the little isolated bubble, my cruising sailboat, told me a lot about the way we use and need energy.

It all started with the lights. There were three 25 watt reading lamps running off a 12 volt storage battery which was hidden up in the stern. The lights came on alright but, after a minute or two, faded to that dim religious light found in expensive restaurants. Not a light for reading by! Plainly some trouble with the battery. Was it charged? Well, 25 watts at 12 volts requires 2 amps. Quite an amount - how long should the battery last? The label said 70 ampere hours - it should last more than a minute or so - even if 35 hours would be rather optimistic.

But that got me thinking and comparing the isolated sailboat with the comforts of home. 70 amp. hours is equivalent to 840 watt hours, or as near as makes no difference when on a boat without a calculator, 1 kwh. Now one pays between 1 and 10 cents for 1 kwh of electricity at home - depending on where you live and all sorts of other things. To get 1 kwh in my boat I have to have a battery worth about $40 so the value of the electricity from the storage device is 100 to 1000 times the value of the electricity when we're ashore.

This of course, is the theme of my little talk - the value of being able to store energy.

I wasn't the only sailor with energy storage problems. On one island we were moored behind a large motor cruiser which had even more comforts of home. One could stand up to cook or even to do macrame in the evening. My daughter was invited to use their hot shower after the mistress of the boat had spotted her rinsing out her hair in a bucket of river water!

The following morning I saw the master of the boat lift up a large hatch at the stern holding, with meaningful air, a pair of battery clips. Curious to see what was down the engine room, I walked up the jetty. The engine was a V-8 of impressive dimensions which didn't seem to be charging properly. Connecting the battery to a second one (needed for some auxiliary machinery) he started up the V-8 and so got his battery charging.

"What's the other battery for", I asked.

"Oh, that's for the generator".

Sure enough - dwarfed by the V-8 was a little 3 kw motor generator! All the comforts of home indeed! Now, 3 kw is equivalent to 30 amps at 100 volts. (Remember one's house is wired for 100 - 200 amps).

Of course, he was very lucky to have the second battery. For me, a dead battery meant we went to bed when it was dark. For him, a dead battery meant that he had an immobile boat. So the $40 for 1 kwh in his battery was very well worth it.

It's interesting to think about the cost of his electricity. The three kw generator must use between half and one gallon of fuel an hour at full load. Sounds like 15-30 cents/kwh for fuel costs alone. But when one's on holiday, who cares about the costs!

When I come to think about it, I had several other energy storage devices on the boat. Storage batteries in a flashlight and a portable radio, a gallon can of methanol for cooking and a 5 gallon gasoline tank for the auxiliary outboard. The flashlight is probably good for 1 watt hour. At 75 cents for batteries that's $750/kwh. See how much we are prepared to pay for stored energy!

The auxiliary outboard is essential for getting a sailboat in and out of docks and harbours, negotiating tricky narrow passages, and of course, for days without wind. For its energy storage there is a 5 gallon gas tank. The tank, like the battery, costs about $40.00. By comparison, however, the 5 gallons of gas are equivalent to about 250 kwh (thermal). That works out to 2 cents/kwh hr. The comparison in storage capacity between the battery and the 5 gallons of gas, which must weigh about the same is worth noting. (That tells you all the basic facts about the electric car). Of course, I don't get 250 kwh of useful work from the outboard engine. Its efficiency is 10-20% so I can get 25 to 50 kwh (mechanical) from the filled tank and so the 6 hp. motor is equivalent to 4.5 kw (mechanical). (These rough figures check. I have about 5 hours cruising on one tank of gas). Now 25 to 50 kwh is really an impressive amount of energy. It explains why hydrocarbon fuels are so valuable; because they provide cheap, convenient and high density energy storage.

Now 4.5 kw. the mechanical energy delivered by the motor is itself quite a chunk of power. It is interesting to compare that with the performance of the sails. Once the wind gets to 10 MPH or so the sails drive the boat just as fast as the motor will. The sails are then equivalent to a 5 kw energy conversion machine. For those who are non-sailors, I must tell you that sails, masts and rigging are not cheap and simple. On this boat, they likely cost well over $1000, possibly near $2000 and so cost two to four times the price of the outboard. They are also, in their way, equally sophisticated pieces of engineering, but that's another story.

The power characteristics of wind and sail differ significantly from an outboard. With wind and sail, one has no throttle control. All the power which can be derived from the sail has to be used which is one of the reasons why sailing is such fun! If the wind increases, the excess power goes to heel the boat. If one gets too much wind, you can release the sails and they flap - but that's only for extreme emergencies and the control of the boat is seriously impaired. It's very poor seamanship. Special devices to control sail area have to be used - or else one puts up special storm sails.

This problem of too much wind is balanced, of course, by the times when there is no wind at all. No wind, no power, and one is stuck - start up the motor! If <u>only</u> one could store the wind from Windy days to put to use on calm days. This rather obvious fact escapes many people nowadays when they talk about wind energy. True it is an old technology, although as I have indicated, by no means a primitive one. But the key to its wide application is energy storage - and cheap energy storage. Remember, even though the wind is free, the sails and masts are quite expensive.

To conclude let me bring all these points together for you - as things to think about when thinking about energy.

(1) Energy is not just one thing. Electrical, mechanical, thermal energy, have their own special values and special uses. We are prepared to pay widely differing amounts of money for different forms of energy.

(2) The ability to store energy is as valuable as energy itself.

(3) Liquid fuels are beautifully convenient and cheap forms of energy storage.

(4) Energy storage is the key to wide application of renewable resources like wind, if they are not to be just supplemental as the sails are in cruising sailboats, and lastly:

(5) Sailing is fun.

AUTHOR INDEX

SUBJECT INDEX

Battery, energy storage, 285

Chemical bond, 43
Composites
  graphite, 223
    fiber reinforced, 257
    reinforced, 239
    residual stresses and strains, 203
    thermal deformation, 203

Density
  in liquid metals and alloys, 97
  of molten metals, 77
Dilatometer
  high temperature, 145
  technique for phase transition, 165

Elastic constants, scandium, 57
Energy storage, 285

Gamma attenuation in molten materials, 83
Graphite composites, 223

Interferometer, 173
Interferometric technique for thermal expansion measurements, 131

Lattice
  thermal expansion, 107
  vibrations, 107

Nuclear
  fuel elements, 1
  power generation, 1
  reactor, 1
    advanced concepts, 1

Phase transition, 165
Priest interferometer, 173

Satellite structures, 223
Scandium
  thermal expansion, 57
  third-order elastic constants, 57
Space optics, materials application, 27
Structure, ABO_4 compounds, 46
Structures, satellite, 223

Thermal expansion, 43
  anomaly in sodium chlorate, 99
  by modulation method, 155
  calcite type crystals, 53
  filament wound composite tubes, 275
  in liquid metals and alloys, 97
  KDP type crystals, 47
  measurement by gamma attenuation, 83
    by interferometric technique, 131
    of molten materials, 83
    technique, 111
  of fused silica, 173
  of graphite, 1
  of graphitic composites, 1